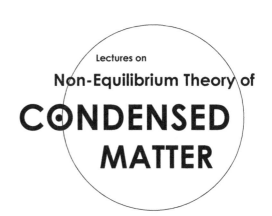

Lectures on

Non-Equilibrium Theory of

CONDENSED
MATTER

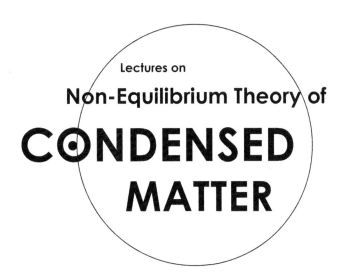

Lectures on
Non-Equilibrium Theory of
CONDENSED MATTER

Ladislaus Alexander Bányai

J W Goethe Universität, Germany

World Scientific

NEW JERSEY · LONDON · SINGAPORE · BEIJING · SHANGHAI · HONG KONG · TAIPEI · CHENNAI

Published by

World Scientific Publishing Co. Pte. Ltd.

5 Toh Tuck Link, Singapore 596224

USA office: 27 Warren Street, Suite 401-402, Hackensack, NJ 07601

UK office: 57 Shelton Street, Covent Garden, London WC2H 9HE

British Library Cataloguing-in-Publication Data

A catalogue record for this book is available from the British Library.

LECTURES ON NON-EQUILIBRIUM THEORY OF CONDENSED MATTER

ISBN-13 978-981-256-749-9
ISBN-10 981-256-749-6

Printed in Singapore

To my beloved wife Anikó for the wonderful life we spent together

Preface

This book intends to describe the problems and methods of non-equilibrium statistical mechanics of condensed matter unto its most recent developments as observed by the author during his scientific career since the early sixties. On the one hand, most topics are from the author's research interests, therefore his choice of these topics is personal to some extent. On the other hand, he would like to emphasize to the reader that this book is not just a pile of simple external compilations.

The content of these lectures was partly scattered in various conferences, seminar talks and lectures held over the years, and partly elaborated for a series of lectures for graduate students and postdocs at the University of Frankfurt in 2004.

The author would like to thank his old friend and coworker Paul Gartner for the years of harmonious collaboration and many exciting debates on the problems of statistical mechanics. He wishes to express his gratitude to Hartmut Haug, for the possibility to collaborate successfully on the most modern topics of non-equilibrium theory. Also, he would like to acknowledge Peter Kopietz for the invitation to give a series of lectures on non-equilibrium many-body theory, which finally prompted the work on this book. Last but not least, many thanks to several dear colleagues for the fruitful cooperation — among them, mostly to Stephan Koch. The author is especially indebted to János Hajdu for the careful reading of the manuscript and the many improvements suggested by him, which were included in the final text, partly as footnotes.

L. Bányai
Frankfurt 2006

Contents

Chapter 1

Introduction (Why and how non-equilibrium?)

Nowadays most theorists working on condensed matter or statistical mechanics are involved with problems related to equilibrium. The interest is focused on different phase transitions. Many of them are spectacular, being candidates for technological applications too. On the other hand, the situation is very clear from a conceptual point of view. The object of mathematical interest is the statistical sum in the thermodynamic limit. However, one has to face a very intricate question - how equilibrium may be reached by physical systems obeying time reversal invariant (classical- or quantum-) mechanics? This is the central problem non-equilibrium theorists have to deal with, and the issue as such is getting a promising yet not satisfactory and exhaustive answer in the last decades only. Although Boltzmann formulated his famous and very successful equation for gases at the end of the 19th century, he was desperate about the lack of any sound foundation, neither was he able to defend it against the harsh criticism of Mach.

Beyond the conceptual problem of evolution versus equilibrium, one was interested in the past mostly in the description of stationary or slow non-equilibrium states, especially in the elaboration of procedures, to calculate on the basis of microscopical models the various phenomenological material parameters. This aspect had a serious motivation, also due to the technological applications based on the properties of metals and semiconductors.

However, the focus soon turned to non-linear phenomena, that proved to be extremely successful in technological applications. Here already one had to concentrate on behaviors and laws, resulting from the governing equations, rather than on a few phenomenological coefficients.

The last kick was given a decade ago by the use of extremely intense and ultrashort laser pulses (with wave-length corresponding to the gaps of most important semiconductor materials). Here a new sort of experiments appeared under the name of "ultrashort-time spectroscopy". In such an experiment one irradiates the material with two or more more ultrashort (in femtosecond domain) light pulses, and this way one is able to control the delay between the pulses too, which is also in the femtosecond range. This way, one gets a new control parameter which can give an insight into the rapid time evolution occurring in a many-body system.

The inventory of the theorist has improved in the past hundred years too. Until the late fifties of the past century a typical textbook on statistical mechanics described the Boltzmann equation as the unique topic of non-equilibrium. The Master equations themselves were treated in a rather marginal manner, since their modern applications seemed very limited.

Ever since then, the linear response theory- related mostly to the name of Kubo - made a breakthrough in the theoretical understanding of the kinetic coefficients, and found its way into the textbooks. The very last evolution in the theory occurred recently, due to the efforts to understand the new insight offered by the ultra-short spectroscopy. This implied the use of modern mathematical methods in treating interacting many-body systems, and tries to answer questions related to the very rapid microscopical changes in such a system. A new feature of the so called "quantum kinetics" is that one has to describe not just a parameter, which is independent on the way to measure it. The new sort of approach is to describe a definite experiment, and it can be formulated only by a close cooperation between the theorist and experimentalist. Although this approach proved to be very fruitful, also, unfortunately made the field less transparent, even if some monographs were already published on the subject [Haug and Jauho (1995); Bonitz (1998)]. Many theorists, however, are still reluctant to face the basic question: what has their calculation to do with the laboratory experiment? Are they calculating the same entity as the one measured in experiment *whether the calculated curves fit or not* ?

This book tries a somewhat unconventional approach. It starts from analyzing some exactly solvable classical models, in order to illustrate the basic paradigms. Afterwards we sketch the approximative quantum-mechanical approach in a simplified way, showing its main features, while we discuss them later on a more sophisticated level. After this general glimpse, we will present the properties of the traditional equations of non-equilibrium statistical mechanics, such as the Master and rate equations as well as the Boltzmann equation. A relatively wide discussion of the scaling connection between the kinetic and hydrodynamic levels follows afterwards.

Several chapters are devoted to linear response theory and its applications. A careful discussion of non-relativistic quantum electrodynamics - as well as of its relationship to macroscopical electrodynamics - fits in this context. The second half of the book is devoted to the more modern aspects, on a more sophisticated level. The Markovian equations for the density operator (including its non-diagonal elements) and their deductions, are described in some detail. We also discuss the Bose-Einstein condensation in real time as an example for a better comprehension of spontaneous symmetry breaking. The last part is devoted to the presentation of the so called quantum kinetics, its mathematical tools (Keldysh-Green functions), properties of the approximations as well as the successes in explaining modern ultra-fast spectroscopy experiments.

Chapter 2

Time-reversal and evolution toward equilibrium

The Lagrange-function of a classical system of point-like interacting particles

$$L = \sum_{i=1}^{N} \frac{m\dot{x}_i^2}{2} - U(x_i) - \frac{1}{2} \sum_{i=1}^{N} \sum_{j=1}^{N} V(x_i - x_j)$$

is invariant against the time-reversal $t \Rightarrow -t$ implying of course the reversal of the velocities $\dot{x} \Rightarrow -\dot{x}$ too. This property is a consequence of the fact that we have to deal with forces deriving from a potential. As a consequence, from any final state one may get back the initial state by reversing the sign of the velocities, i.e. the system is *reversible*.

There is a special case of velocity dependent forces, namely the electromagnetic ones, which, by a slight extension, may be also incorporated into the Hamilton-Lagrange formalism. In this case one has to inverse the sign of the magnetic field too, in order to recover the initial state.

Poincaré demonstrated in this context that after a definite time, every mechanical system returns to the vicinity of the initial state in the phase space.

Within quantum mechanics, time-reversal stems from the reality of the Hamiltonian operator $H^* = H$, from which it follows that if $\psi(t)$ is a solution of the Schrödinger equation

$$\imath\hbar\frac{\partial}{\partial t}\psi = H\psi \,,$$

then $\psi(-t)^*$ is also a solution. (In the presence of a magnetic field one has again to reverse the magnetic field too.)

This seems impossible to reconcile with the observed irreversibility of the physical phenomena and the tendency of physical systems towards equilibrium. Indeed, from a unique final equilibrium state it is impossible to return to the infinite manifold of initial states.

One of the simplest phenomenological laws in working with electricity is Ohm's law

$$\vec{j} = \sigma \vec{E},$$

relating the current density - flowing in a conductor - to the intensity of the applied field. This law obviously shows a violation of the time-reversal invariance. The current behaves by time reversal like a velocity (changes its sign), while the electric field, being a force, behaves like an acceleration (i.e. remains unchanged). Hence, this law cannot be understood within the frame of mechanics (or quantum mechanics). Indeed, according to the Newtonian equation of motion, the first time-derivative of the velocity should be proportional to the field and not to the velocity itself. The motion of an electron in a constant electric field with a constant velocity may be explained only by the presence of phenomenological "non-mechanical forces". If one introduces a damping term with the damping constant κ in Newton's equation of motion

$$m\ddot{x}(t) = eE - \kappa\dot{x}(t) \tag{2.1}$$

then, indeed -starting from any initial state - the asymptotic motion is that with stationary velocity

$$\dot{x} \to \frac{eE}{\kappa}.$$

The paradox cannot be solved within the mechanics of closed systems. However, one should realize that real systems are never closed: they always interact with the surrounding. Since we omit the degrees of freedom of the surrounding, they have to be considered as subsystems, which cannot have a pure mechanical description. A mechanical description may be good on short time scales - (depending on the strength of the interaction with the surrounding) - however, this interaction cannot be ignored on a large time scale. On the other hand, one may ignore the changes of the surrounding itself only when the surrounding is very large.

Therefore, a basic ingredient of the study of non-equilibrium systems is the consideration of open systems with an infinite number of degrees of freedom, in an infinitely large volume at a finite density. This condition is a must at least for one component of the system (surrounding). When in thermal equilibrium, then it is called a thermostat.

The thermostat plays an essential role in the evolution of an open system towards equilibrium, as well as in its reaction to external perturbations (like the applied field in the charge transport). As we shall see later, the distinction between the thermostat and the system itself may be very subtle, pertinent to certain degrees of freedom.

In order to have a clear understanding of the paradigm, we shall analyze two exactly solvable classical models for the motion of an electron in an electric field in the following two chapters. We consider it a necessary step, as the complexity of realistic descriptions hinders the full transparency of the underlying assumptions in their treatment.

There is a class of systems (superconductive or superfluid), which remain frictionless under a critical temperature, even if in contact with a thermal reservoir. These special states of the matter obey time-reversible laws. For example, in a superconductor, the time derivative of the current is proportional to the electric field. While the equilibrium properties of these phenomena may be well understood, a satisfactory theory of their non-equilibrium properties - i.e. their stability against breaking time-reversal - still remains an utmost difficult question. In these lectures we are not going to dwell on these special states of the matter.

Chapter 3

The 1D model of Caldeira and Leggett

Let us consider a classical one-dimensional model [Caldeira and Leggett (1983)] for the motion of an electron (of unit mass and unit charge) of coordinate x, interacting with acoustical phonons in the presence of a constant electric field $\mathcal{E}$. The acoustical phonons are described by the oscillator coordinates Q_q for each wave-vector q having the oscillator frequencies $\omega_q = c|q|$, with c being the sound velocity. The continuous wave-vectors of phonons are cut-off at the Debye wave-vector q_D. The Lagrangian of the whole system is

$$\mathcal{L} = \frac{1}{2}\dot{x}^2 + \mathcal{E}x + \frac{1}{2}\int_{|q|<q_D} dq \left\{ \dot{Q}_q^2 - \left(\frac{F_q}{\omega_q}x - \omega_q Q_q \right)^2 \right\}. \qquad (3.1)$$

The interaction consists in a shift of the oscillator potential proportional to the coordinate of the electron, and we consider here

$$F_q = \sqrt{\frac{c\kappa}{\pi}}\omega_q \qquad (3.2)$$

where κ is a new constant, chosen for convenience.

The coupled Euler-Lagrange equations for the phonons and the electron are

$$\ddot{Q}_q = \omega_q \left(\frac{F_q}{\omega_q}x - \omega_q Q_q \right) \qquad (3.3)$$

$$\ddot{x} = \mathcal{E} - \int_{|q|<q_D} dq \frac{F_q}{\omega_q} \left(\frac{F_q}{\omega_q}x - \omega_q Q_q \right)$$

$$= \mathcal{E} - \int_{|q|<q_D} dq \frac{F_q}{\omega_q^2} \ddot{Q}_q .$$

9

One chooses initial conditions corresponding to no phonon ellongations or oscillations at $t = 0$, while the electron is at the origin with an arbitrary initial velocity $\dot{x}(0)$

$$Q_q(0) = 0; \quad \dot{Q}_q(0) = 0; \quad x(0) = 0.$$

The formal solution of the first of these equations for the phonon variables, in terms of the electron variable (trajectory), is

$$Q_q(t) = \frac{F_q}{\omega_q} \int_0^t dt' x(t') \sin \omega_q(t - t')$$

Taking the second derivative of this equation followed by partial integrations one gets

$$\frac{d^2}{dt^2} Q_q = F_q \frac{d}{dt} \int_0^t dt' x(t') \cos \omega_q(t - t')$$

$$= F_q \left(x(t) + \int_0^t dt' x(t') \frac{d}{dt} \cos \omega_q(t - t') \right)$$

$$= F_q \left(\int_0^t dt' \dot{x}(t') \cos \omega_q(t - t') + x(0) \cos \omega_q t \right),$$

and with $x(0) = 0$ one gets

$$\frac{d^2}{dt^2} Q_q = F_q \int_0^t dt' \dot{x}(t') \cos \omega_q(t - t').$$

Introducing this result in the last of the Eqs.(3.4) one gets

$$\ddot{x}(t) = \mathcal{E} - \int_{|q| < q_D} dq \frac{F_q^2}{\omega_q^2} \int_0^t dt' \dot{x}(t') \cos \omega_q(t - t').$$

Implementing now our choice of F_q

$$\ddot{x}(t) = \mathcal{E} - \frac{2}{\pi} \kappa \int_{0 < \omega_q < \omega_D} d\omega_q \int_0^t dt' \dot{x}(t') \cos \omega_q(t - t'), \tag{3.4}$$

and performing the integration over the frequency one gets the final closed, but non-local (in time) equation for the electron coordinate

$$\ddot{x}(t) = \mathcal{E} - \frac{2}{\pi} \kappa \int_0^t dt' \frac{\sin \omega_D(t - t')}{t - t'} \dot{x}(t'). \tag{3.5}$$

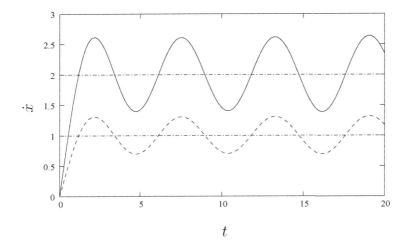

Fig. 3.1 Numerical solution of the Caldeira-Legget model for $\frac{\mathcal{E}}{\kappa} = 1, 2$ - the time being in natural units $\frac{1}{\omega_D}$.

This non-locality (memory) is the price we had to pay for the elimination of the phonon variables. Remark that due to the integration over the continuous phonon frequency one got finally a kernel $\frac{\sin \omega_D (t-t')}{t-t'}$ that is concentrated at $t = t'$.

In Fig.(3.1) the numerical solution of this equation is shown for two choices of the field intensity $\frac{\mathcal{E}}{\kappa} = 1, 2$. The time is given in natural units $\frac{1}{\omega_D}$. At the beginning the velocity increases linearly, as it is expected for an accelerated electron in electric field, but for large times it is oscillating around $\frac{\mathcal{E}}{\kappa}$ (here 1, respectively 2). This may be easily understood by checking that since

$$\int_{-\infty}^{0} du \frac{\sin u}{u} = \frac{\pi}{2} ,$$

the constant velocity

$$\dot{x} = \frac{\mathcal{E}}{\kappa}$$

satisfies the asymptotic limit of Eq. (3.5), i.e. the one with the upper limit of the integration extended to infinity.

Leaving apart the oscillations (with the frequency ω_D), this is exactly the kind of miraculous Ohmic behavior we discussed in the previous chapter contradicting time-reversal invariance. Moreover, one may say that the slowly varying, smooth component of the velocity satisfies the equation of motion with the simple Drude damping like in Eq.(2.1). After the presentation of another solvable model in the next chapter we comment on the things we learned.

Chapter 4

The classical polaron in a d.c. field

In this chapter we consider another solvable classical model [Bányai (1993)] - now in the three-dimensional space - describing an electron (of unit mass and charge) moving in a homogeneous electric field and interacting with LO phonons (of unit frequency). This model is nothing but the classical version of the Fröhlich interaction, and it is described by the Lagrange function

$$
L = \frac{1}{2}\dot{\vec{r}}^2 + \vec{\mathcal{E}}\vec{r} + \int d\vec{x} \left\{ \dot{\vec{u}}(\vec{x})^2 - \vec{u}(\vec{x})^2 + \sqrt{\frac{C}{4\pi}} \nabla \vec{u}(\vec{x}) V(\vec{x} - \vec{r}) \right\} . \quad (4.1)
$$

Here $\vec{r}$ is the coordinate of the electron, $\vec{u}(\vec{x})$ describes the local ionic displacement, C is a coupling constant, and $\vec{\mathcal{E}}$ is the d.c. electric field.

The Euler-Lagrange equations for the electron coordinate and for the longitudinal phonon degree of freedom

$$
d(\vec{x}, t) \equiv \nabla \vec{u}(\vec{x}, t) , \quad (4.2)
$$

which is the only one coupled to the electron are

$$
\ddot{d}(\vec{x}, t) + d(\vec{x}, t) = -\sqrt{\frac{C}{4\pi}} \nabla^2 V(\vec{x} - \vec{r}(t)) \quad (4.3)
$$

$$
\ddot{\vec{r}}(t) = -\sqrt{\frac{C}{4\pi}} \int d\vec{x} d(\vec{x}, t) \nabla V(\vec{x} - \vec{r}(t)) + \vec{\mathcal{E}}. \quad (4.4)
$$

We chose again an initial state at $t = 0$ with the electron in the origin and without phonons

$$d(\vec{x}, 0) = 0 \qquad \dot{d}(\vec{x}, 0) = 0 \,. \tag{4.5}$$

As in the previous chapter, we may solve formally the phonon equation expressing $d(\vec{x}, t)$ through the electron trajectory

$$d(\vec{x}, t) = -\sqrt{\frac{C}{4\pi}} \int_0^t dt' \sin(t - t') \nabla^2 V(\vec{x} - \vec{r}(t')) \,, \tag{4.6}$$

and introducing it in the equation for the electron coordinate we get

$$\ddot{\vec{r}}(t) = \vec{\mathcal{E}} + C \int_0^t dt' \sin(t - t') \frac{\partial}{\partial \vec{r}(t)} W(\vec{r}(t) - \vec{r}(t')) \,. \tag{4.7}$$

Here we have introduced the notation

$$W(\vec{r}) \equiv -\frac{1}{4\pi} \int d\vec{x} V(\vec{r} - \vec{x}) \nabla^2 V(\vec{x}) \,. \tag{4.8}$$

In the following we restrict ourselves to motions along the field direction. Then we have again a one-dimensional equation

$$\frac{d^2 x(t)}{dt^2} = \mathcal{E} + C \int_0^t dt' \sin(t - t') \frac{\partial}{\partial x(t)} W(|\, x(t) - x(t')\,|) \,, \tag{4.9}$$

or

$$\frac{d^2 x(t)}{dt^2} = \mathcal{E} - C \int_0^t dt' \sin(t - t') K(x(t) - x(t'))$$

with

$$K(x) = -\frac{\partial}{\partial x} W(|\, x\,|) \,.$$

Ideal would be to consider here the Coulomb potential $V(\vec{x}) = \frac{1}{|\vec{x}|}$ in order to get the classical version of the Fröhlich coupling to LO phonons. However again, due to the fact that the continuum model for phonons introduces a dangerous artefact, one must limit the number of the phonon degrees of freedom (otherwise the model is mathematically ill-defined); this implicitly modifies the potential, since an upper cut-off of the coupled phonon modes in q-space (as in the Caldera-Leggett model) is equivalent to a lower cut-off of the potential in the x-space. We choose the potential as

$$V(r) = \frac{1}{r}\left(1 - e^{-r}\right) . \tag{4.10}$$

This potential behaves at large distances as a Coulomb one, but has a finite value in the origin (see Fig. (4.10)). By this choice we have fixed the length unit to the cutoff distance.

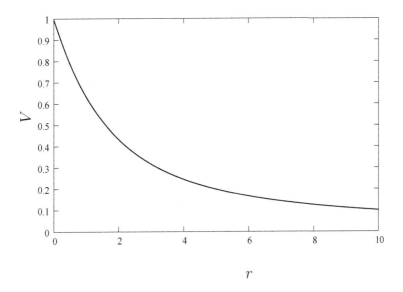

Fig. 4.1   The potential $V(r)$.

With this choice

$$W(r) = \frac{1 - e^{-r}}{r} - \frac{1}{2}e^{-r} ,$$

and

$$K(x) = -sign(x)\left(\frac{e^{-x}}{2} - \frac{1 - e^{-x}}{x^2} + \frac{e^{-x}}{x}\right) .$$

The kernel $K(x)$ is represented in Fig. (4.2).

Let's look at the beginning for solutions describing an asymptotic uniform motion with the velocity v

$$x(t) \approx vt \qquad \text{for} \qquad t \to \infty . \tag{4.11}$$

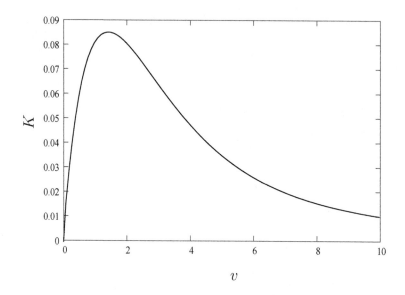

Fig. 4.2　The kernel $K(x)$.

Inserting into the asymptotic form of the equation (with the integration range extended to infinity) we have

$$0 = \mathcal{E} - C \int_0^\infty dt \sin t K(vt) \, . \tag{4.12}$$

Performing the integral we get a transcendent equation, which restricts the possibilities for such a solution

$$\frac{\mathcal{E}}{C} = H(v) \, , \tag{4.13}$$

with the characteristic function

$$H(v) = -\frac{1}{2\left(1 + v^2\right)} + \frac{\ln\left(1 + v^2\right)}{2v^2} \, .$$

This function is shown on Fig.(4.3). Its main implication is that for every ratio $\frac{\mathcal{E}}{C}$ there is a critical velocity, above which no asymptotic stationary solution exists. This can be better seen on Fig. (4.4), where the allowed stationary velocities are shown as function of $\frac{\mathcal{E}}{C}$. Then it shows a critical $\frac{\mathcal{E}}{C}$.

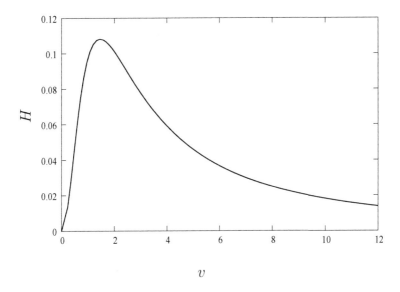

Fig. 4.3 Characteristic function $H(v)$.

As it may be seen from the numerical calculations, not all the stationary velocities allowed by this curve are truly realized. Only the ones on the lower branch are stable solutions, and even in those cases the solution often oscillates around the predicted value, as in the case of the Caldera-Leggett model.

Another important implication of the shape of the allowed velocities is that the resulting velocity-field relationship is - generally speaking - not a linear one as in the Caldera-Legget model. In other words, the stationary current-field characteristic is not that given by the simple Ohm law.

Now let us see some typical trajectories obtained by numerical solution of Eq.(4.9).

In Fig.(4.5) a typical dissipative asymptotic solution is shown. The velocity increases at the beginning, and after some transient oscillations approaches his asymptotic stationary value, which coincides with the one predicted by the corresponding curve of Fig.(4.4).

Above criticality the asymptotic behavior is purely ballistic, the electron gets accelerated by the field as it is shown on Fig. (4.6). This later behavior is a typical "breakthrough" phenomenon, known also in the physics of dielectrics. A similar "breakthrough" may occur even at fields much below the critical one, if the initial velocity is taken bigger as the critical one.

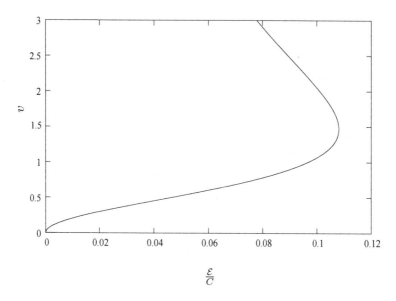

Fig. 4.4 Asymptotic velocity as function of $\mathcal{E}/C$.

The general conclusion from the numerical studies of this classical model is that in the presence of a d.c. electric field only two kind of asymptotic motions are possible, depending on the range of parameters: i) Dissipative motions with constant velocity. (*However, with a nonlinear dependency on the field and sometimes also oscillating.*) ii) Ballistic motions with constant acceleration identical to that of free electrons. Of course, even in the case of the dissipative motion the energy is conserved, but it is being transfered to the phonons.

An interesting question regarding this model is whether the velocity of a moving electron in the absence of an electric field is damped, and whether this damping occurs according to an exponential decay as a phenomenological Drude damping term $-\kappa \dot{x}(t)$ in the Newton equation would produce?

A general answer is not at all obvious, however, again one may illustrate the situation by numerical solution. In Fig.(4.7) one may see the damping of the motion started with a finite velocity in the absence of an electric field.

A better answer to this question might come from a local approximation to the Eq.(4.9), which is expected to be valid in the asymptotic regime. The local approximation consist in replacing the non-local term by its asymptotic form given by Eq.(4.3):

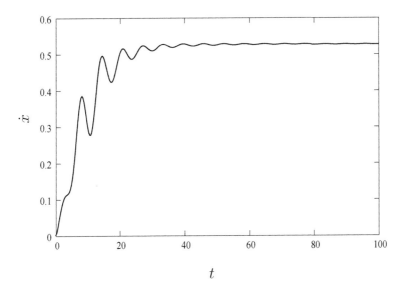

Fig. 4.5   Velocity $\dot{x}(t)$ of a trajectory with $\dot{x}(0) = 0$. ($C = 1$, $\mathcal{E} = 0.05$.)

$$\frac{d}{dt}v(t) = -CH(v(t)) + \mathcal{E}\,, \qquad (4.14)$$

which might describe the slowly varying part (smooth envelope) of the velocity.

This looks as a phenomenological damping

$$\frac{d}{dt}v(t) = -\frac{v(t)}{\tau(v(t))} + \mathcal{E}\,, \qquad (4.15)$$

with a velocity dependent relaxation time $\tau(v)$

$$\frac{1}{\tau(v)} \equiv C\frac{H(v)}{v}\,. \qquad (4.16)$$

For very small velocities we get then the simple equation

$$\frac{d}{dt}v(t) \approx -\frac{C}{4}v(t)^2 \qquad \text{for} \qquad v \to 0\,, \qquad (4.17)$$

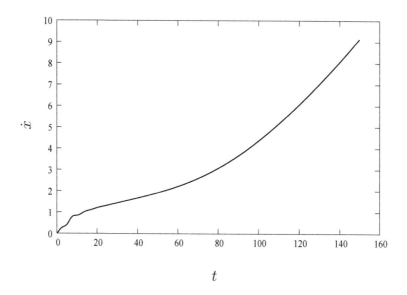

Fig. 4.6   Velocity $\dot{x}(t)$ of a trajectory with $\dot{x}(0) = 0$. ($C = 1$, $\mathcal{E} = 0.13$.)

which may be analytically solved giving rise to

$$v(t) \approx \frac{4}{Ct} \, . \tag{4.18}$$

Therefore, it seems that damping occurs, but only as a slow power law.

On its turn, the correctness of the local approximation can be checked by comparing its solution to numerical solutions of the original non-local equation. We give here some encouraging examples as the one in Figs.(4.8-4.10) for the previously discussed three trajectories. Obviously, the local approximation just smoothes out the true oscillating trajectory, but follows its general shape.

The classical polaron model, like the previously described Caldera-Legget model, describes an electron moving in a d.c. electric field, interacting in the same time with another system (phonons) having an infinite number of degrees of freedom (thermostat). In both cases, after the elimination of the phonon variables, the closed equation for the electron trajectory is non-local in time. However, in order to get the solution for later times one must know the solution for all previous times until the

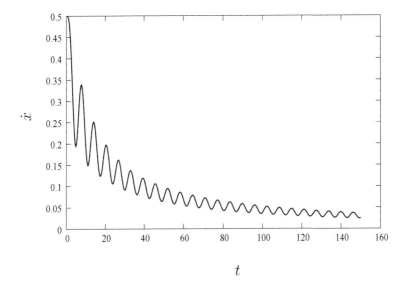

Fig. 4.7   Velocity $\dot{x}(t)$ of a trajectory with $\dot{x}(0) = 0.5$. ($C = 1$, $\mathcal{E} = 0$.)

initial time $t = 0$, where the exact state of the whole system (electrons and phonons) was known. This is a surprising new aspect if we compare it with the standard equations of mathematical physics. Giving the coordinate and velocity of the electron at a later time $t_0 > 0$ is not enough as a new initial condition for the further development. This resulting equation for the electron is irreversible, although for the whole electron-phonon system time-reversal still holds. After having eliminated the infinity of phonon degrees of freedom with a definite initial condition, one cannot reverse the phonon velocities any more, and therefore time-reversal does not work. The asymptotic solution describes either a stationary dissipative motion - with possible oscillations around it - or an accelerated ballistic one without damping. A local approximation, supposedly valid for the slowly varying part of the velocity, follows the global features of the true trajectory, but lacks a lot of rapid details. Within this approximation, one has a velocity-dependent relaxation time.

All these features will be present also in the treatment of more realistic quantum mechanical many-body models, however, due to the enormous

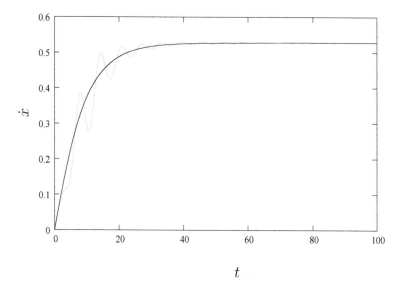

Fig. 4.8 Velocity $\dot{x}(t)$ of a trajectory with $\dot{x}(0) = 0$ ($C = 1$, $\quad \mathcal{E} = 0.05$) calculated within the local approximation, compared to the exact one.

complexity of the problem, an exact solution is not possible; in this case the manifold of approximations make the whole approach less transparent. Therefore, the solvable classical models may serve to justify the basic paradigm of the general approach to non-equilibrium. [1]

---

[1] An instructive example for the elimination of infinitely many degrees of freedom (fields), leading to non-local evolution and breaking of the time reversal symmetry (irreversibility) is the well-known treatment of the electromagnetic self-interaction of a charged particle in the framework of classical electrodynamics (cf. [Jackson (1999)]), although, this self-accelerating electron is a rather unphysical model. Expressing the electromagnetic field produced by the particle in terms of the retarded potentials $V^i(\vec{r}, t) = \int d\vec{r}' \frac{\rho(\vec{r}', t - \frac{R}{c})}{R}$, $\vec{A}^i(\vec{r}, t) = \frac{1}{c} \int d\vec{r}' \frac{\vec{j}(\vec{r}', t - \frac{R}{c})}{R}$, where $\rho(\vec{r}, t)$ and $\vec{j}(\vec{r}, t)$ are the charge and current distributions of the particle respectively, (here $R \equiv |\vec{r} - \vec{r}'|$) and inserting the result into the equation of motion of the particle $m\ddot{\vec{r}} = \vec{F}$ with the Lorentz force $\vec{F}(t) = \int d\vec{r} \left(\rho(\vec{r}, t)\vec{E}^i(\vec{r}, t) + \frac{1}{c}\vec{j}(\vec{r}, t) \times \vec{B}^i(\vec{r}, t) \right)$ we get a non-local (integro-differential) equation for the particle coordinate $\vec{r}(t)$. Expanding now in powers of the retardation $\frac{R}{c}$, and retaining only the first two terms, we end up, in the instantaneous rest frame of the particle, with a Lorentz force proportional to the second derivative of the velocity $\ddot{\vec{v}}$. This self-force term manifestly breaks the time reversal symmetry. The origin of irreversibility lies again in the elimination of the field degrees of freedom by a definite choice of the initial condition for the field (here the retarded one).

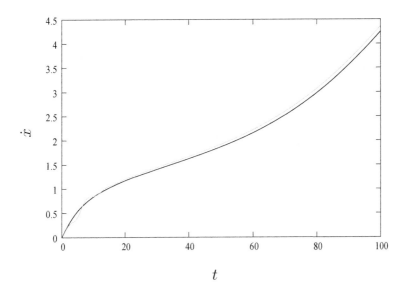

Fig. 4.9   Velocity $\dot{x}(t)$ of a trajectory with $\dot{x}(0) = 0$ ($C = 1,\quad \mathcal{E} = 0.13$) calculated within the local approximation, compared to the exact one.

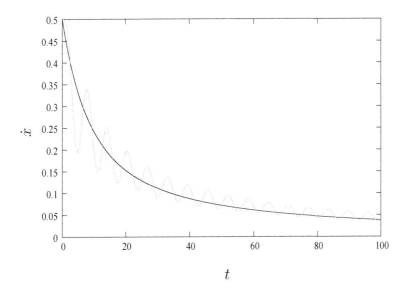

Fig. 4.10   Velocity $\dot{x}(t)$ of a trajectory with $\dot{x}(0) = 0.5$ ($C = 1,\quad \mathcal{E} = 0$) calculated within the local approximation, compared to the exact one.

Chapter 5

Schrödinger equation versus Master equation

Unfortunately, in the study of real quantum-systems one cannot obtain the connection between the ballistic and dissipative behaviors without a chain of non-controllable approximations. Solvable models - as the classical models described in the previous chapters - are not known. On the other hand, it was felt more pragmatical to elaborate general equations describing the evolution towards equilibrium while incorporating elements of quantum mechanics. Indeed, a physicists in his everyday work, is confronted with a dual description of the phenomena. The basic understanding of the nature stems from the Quantum Mechanics of finite, isolated systems (finite number of particles in a finite volume), i.e. the Schrödinger equation for normalized the probability amplitude $\psi(t)$

$$i\hbar \frac{\partial}{\partial t} \psi(t) = H\psi(t) \,,$$

$$\int dx |\psi(x,t)|^2 = 1 \,,$$

which describes reversible evolutions. On the other hand, whenever one has to solve problems related to irreversible macroscopical phenomena, needs equations which perhaps are not lying on a very sound basis, but which mix microscopical elements together with the ingredient of irreversibility. Among these equations the Master equations (and the closely related rate equations) are perhaps the most transparent and successful. A Master equation for the probability of a certain quantum-mechanical state k expresses just the loss and gain balance due to transitions between the unperturbed states of the system caused by the interaction (usually with a reservoir)

$$\frac{\partial}{\partial t}p_k(t) = -\sum_{k'}\{W_{k,k'}p_k(t) - W_{k',k}p_{k'}(t)\}\,;\qquad(5.1)$$

$$\sum_k p_k(t) = 1\,.$$

The transition rates one admits to be given by the "golden rule" of Quantum Mechanics, in which an averaging over the initial state of the reservoir (thermostat) is performed:

$$W_{k,k'} = \frac{2\pi}{\hbar}\sum_{\alpha,\alpha'}\mathcal{P}_\alpha|\langle k,\alpha|H_{int}|k',\alpha'\rangle|^2\delta(\epsilon_k + E_\alpha - \epsilon_{k'} - E_{\alpha'})\,.\qquad(5.2)$$

Here $\mathcal{P}_\alpha$ is the probability of a given reservoir-state α, and of course the interaction Hamiltonian H_{int} between the system and the reservoir produces the transitions.

This equation, as we analyze later, indeed, leads to thermal equilibrium for the system. Also, from the very way it is established, this implementation of quantum mechanical elements in a statistical picture seems to be very natural and logical. Unfortunately, there is a built-in paradox.

The initial condition

$$p_k(0) = \delta_{k,k_0}$$

implies the identification of the derivative of the probability at $t = 0$ with the transition rate itself

$$\frac{\partial}{\partial t}p_k(t)|_{t=0} = W_{k,k_0}\ for\ k \neq k_0\,.$$

On the other hand, the textbook derivation of the "golden rule" starts from the calculation of the probability amplitude $c_{k,\alpha}(t)$ of a system-reservoir state k, α taken to be initially in the state k_0, α_0, i.e.

$$c_{k,\alpha}(0) = \delta_{k,k_0}\delta_{\alpha,\alpha_0}\,.$$

Then the transition rate from the state k_0, α_0 to the state k, α is the asymptotic limit of the rate of the probability :

$$W_{k,\alpha,k_0,\alpha_0} = \lim_{t\to\infty}\frac{\partial}{\partial t}|c_{k,\alpha}(t)|^2\,.$$

If one takes an average over the initial states of the thermostat with the probability $\mathcal{P}_{\alpha_0}$, and sums over all the possible final states α of the thermostat

$$\sum_{\alpha,\alpha_0} \mathcal{P}_{\alpha_0} W_{k,\alpha,k_0,\alpha_0} \,,$$

gets the transition rate W_{k,k_0}, which in the lowest order adiabatic perturbation theory leads to Eq.(5.2). Since the probability p_k itself is defined by

$$p_k(t) = \sum_{\alpha,\alpha'} \mathcal{P}_\alpha |c_{k,\alpha}(t)|^2 \,,$$

we encounter a dilemma: does the transition rate W_{k,k_0} coincide with the time derivative of the probability $p_k(t)$ at $t = 0$ - as the Master equation dictates - or it coincides with the time derivative of the probability at $t = \infty$ as the quantum-mechanical derivation it requires? The two statements contradict each other. The situation suggests that for the validity of the Master equation the time should be short enough on a certain scale, but also large enough on another scale.

The situation may be illustrated also by the following simulation of a Master equation by a "Gedanken-experiment" thought as successive steps of quantum-mechanical evolutions and complete measurements. Here the measurements introduce the opening of the system. (Pauli formulated it as a hypothesis of random phases [Pauli (1928)].) In this case, no interaction with a thermostat is introduced.

Let us consider at time t a mixed ensemble of states $|n\rangle$ with the probabilities $P_n(t)$. Now, let the system develop according to quantum mechanics until the time $t + \Delta t$. At this new time with the same probabilities one has the states $|n(t)\rangle = U(t, t + \Delta t)|n\rangle$, where $U(t, t + \Delta t)$ is the unitary evolution operator. The probability of a state $|n'\rangle$ is now $|\langle n'|U(t, t + \Delta t)|n\rangle|^2$. If one ignores the coherent part of the information about the system, or equivalently, after a complete measurement of the observable n gets the new mixed ensemble

$$P_n(t + \Delta t) = \sum_{n'} P_{n'}(t)|\langle n'|U(t, t + \Delta t)|n\rangle|^2 \,,$$

or

$$P_n(t + \Delta t) - P_n(t) = \sum_{n'} \{ |\langle n'|U(t, t + \Delta t)|n\rangle|^2 P_{n'}(t)$$
$$- |\langle n|U(t, t + \Delta t)|n'\rangle|^2 P_n(t) \} ,$$

where we used the unitarity of the quantum-mechanical evolution and the completeness of the states.

The Master equation with

$$W_{nn'} = \frac{1}{\Delta t} |\langle n'|U(t, t + \Delta t)|n\rangle|^2 , \tag{5.3}$$

emerges if Δt is short enough to approximate

$$P_n(t + \Delta t) - P_n(t) \approx \frac{d}{dt} P_n(t) \Delta t$$

whereas in the same time Δt is long enough to approximate

$$|\langle n'|U(t, t + \Delta t)|n\rangle|^2 \approx \Delta t \lim_{\tau \to \infty} \frac{\partial}{\partial \tau} |\langle n'|U(t, t + \tau)|n\rangle|^2 ,$$

instead of

$$|\langle n'|U(t, t + \Delta t)|n\rangle|^2 \approx \Delta t \lim_{\tau \to 0} \frac{\partial}{\partial \tau} |\langle n'|U(t, t + \tau)|n\rangle|^2 .$$

We will see later by the modern mathematical derivation of the Master equation, the way these ideas about short and long get mathematically implemented.

Chapter 6

Naive deduction of the rate equation from the equation of motion

Although we discuss later in detail the quantum-mechanical derivation of the Master and rate equations, we are going to give here a simplified version for the derivation of the rate equation so often found in the literature. (The relationship in itself between the Master and rate equations will be discussed in a later chapter.) This kind of derivation (which I called here "naive") just indicates which approximations have to be implemented in order to obtain the desired result, without any deep justification of the approximations themselves.

Let us consider a 2-level fermion system coupled to a phonon bath with a continuous spectrum. The Hamiltonian of the whole system is

$$H = \epsilon_1 a_1^\dagger a_1 + \epsilon_2 a_2^\dagger a_2 + \sum_\alpha \{\hbar \omega_\alpha b_\alpha^\dagger b_\alpha + g_\alpha (a_2^\dagger a_1 b_\alpha + a_1^\dagger a_2 b_\alpha^\dagger)\}.$$

Now let us consider the average occupation of the states:

$$f_n \equiv \langle a_n^\dagger a_n \rangle = Tr\{\rho a_n^\dagger a_n\}, \quad (n = 1, 2),$$

where the averaging is with the density operator ρ describing the quantum-statistical state.

From the Heisenberg equation of motion for the operator $a_1^\dagger a_1$ one gets

$$i\hbar \frac{\partial}{\partial t} f_1 = -\langle [H, a_1^\dagger a_1] \rangle$$
$$= \sum_\alpha g_\alpha (\langle a_2^\dagger a_1 b_\alpha \rangle_t - \langle a_1^\dagger a_2 b_\alpha^\dagger \rangle_t).$$

Obviously, this equation does not close for the occupation numbers; the right hand side of the last equation brings new averages in the game. One may write down also the equations for the new averages:

$$i\hbar \frac{\partial \langle a_2^\dagger a_1 b_\alpha \rangle}{\partial t}|_{coll} = -\langle [H, a_2^\dagger a_1 b_\alpha] \rangle$$

$$= (\epsilon_2 - \epsilon_1 - \hbar\omega_\alpha)\langle a_2^\dagger a_1 b_\alpha \rangle$$

$$+ \sum_{\alpha'} g_{\alpha'} \langle (a_2^\dagger a_1 b_{\alpha'} + a_1^\dagger a_2 b_{\alpha'}^\dagger), a_2^\dagger a_1 b_\alpha \rangle .$$

We see that by using again the Heisenberg equation of motion: new more complicated averages appear, like $\langle a_1^\dagger a_1 a_2 a_2^\dagger b_\alpha b_{\alpha'}^\dagger \rangle$. In order to break this seemingly endless game, one tries to implement a simple approximation, decoupling the higher averages (with more operators) into lower ones

$$\langle a_1^\dagger a_1 a_2 a_2^\dagger b_\alpha b_{\alpha'}^\dagger \rangle \approx \langle a_1^\dagger a_1 \rangle \langle a_2 a_2^\dagger \rangle \langle b_\alpha b_\alpha^\dagger \rangle \delta_{\alpha,\alpha'} ,$$

or even more drastically, replaces the average occupation number of the phonons $\langle b_\alpha b_\alpha^\dagger \rangle$ by their equilibrium values, given by the Bose distribution $N_\alpha = \frac{1}{e^{\beta\hbar\omega}-1}$ getting

$$\langle a_1^\dagger a_1 a_2 a_2^\dagger b_\alpha b_{\alpha'}^\dagger \rangle \approx f_1(1 - f_2)(N_\alpha + 1)\delta_{\alpha,\alpha'} . \tag{6.1}$$

Surely, this simplification ignored some of the correlations induced by the interaction. Now we may solve formally the equation for the new variables getting

$$\langle a_2^\dagger a_1 b_\alpha \rangle - \langle a_1^\dagger a_2 b_\alpha^\dagger \rangle = -\frac{2ig_\alpha}{\hbar} \Re \int_{t_0}^t dt' e^{i(\epsilon_2 - \epsilon_1 - \hbar\omega_0)\frac{(t-t')}{\hbar}}$$

$$\times \{f_1(t')(1 - f_2(t'))(N_\alpha + 1)$$

$$- f_2(t')(1 - f_1(t'))N_\alpha\} .$$

Here we took vanishing initial conditions for the new variables at $t = t_0$ (i.e. no initial correlations). After its insertion in the equation of f_1 we get

$$\frac{\partial}{\partial t} f_1(t) = -\frac{2}{\hbar^2} \sum_\alpha g_\alpha^2 \int_{t_0}^t dt' \cos((\epsilon_2 - \epsilon_1 - \hbar\omega)\frac{(t - t')}{\hbar})$$

$$\times \{f_1(t')(1 - f_2(t'))(N_\alpha + 1) - f_2(t')(1 - f_1(t'))N_\alpha\} .$$

Now, let us denote by $M(\omega)$ the thermodynamic limit for the phonons of the expression

$$\frac{1}{\hbar}\sum_\alpha g_\alpha^2 \delta(\omega_\alpha - \omega).$$

Then, taking into account also that the Bose distribution depends only on the energy, we get

$$\frac{\partial}{\partial t}f_1(t) = -\frac{2}{\hbar}\int d\omega M(\omega)\int_{t_0}^t dt' \cos((\epsilon_2 - \epsilon_1 - \hbar\omega)\frac{(t-t')}{\hbar})$$
$$\times \{f_1(t')(1 - f_2(t'))(\mathcal{N}(\hbar\omega) + 1) - f_2(t')(1 - f_1(t'))\mathcal{N}(\hbar\omega)\}.$$

This resembles the rate equation, but it is still an equation with memory. The right hand side of the equation depends explicitly on the values of f_1 and f_2 at earlier times. According to the basic property of the Fourier transforms, the depth of the memory is roughly speaking inversely proportional to the spectral width of the phonon spectrum, i.e. the spectral width of $M(\omega)$.

Now, for very large times $t \to \infty$, admittedly $f_i(t')$ and $f_2(t')$ vary slowly during the memory depth and one may take it out from under the time integration with its value at $t' = t$

$$\int d\omega M(\omega)\int_{t_0}^t dt' \cos\left((\epsilon_2 - \epsilon_1 - \hbar\omega)\frac{(t-t')}{\hbar}\right)f_1(t')(1 - f_2(t'))$$
$$\approx f_1(t)(1 - f_2(t))\int d\omega M(\omega)\int_0^\infty dt' \cos\left((\epsilon_2 - \epsilon_1 - \hbar\omega)\frac{t'}{\hbar}\right)$$
$$= \pi f_1(t)(1 - f_2(t))\int d\omega M(\omega)\delta(\epsilon_2 - \epsilon_1 - \hbar\omega)$$
$$= \pi f_1(t)(1 - f_2(t))M(\epsilon_2 - \epsilon_1).$$

Then indeed we get the rate equation

$$\frac{\partial}{\partial t}f_1(t) = -W\{f_1(t)(1 - f_2(t))(N + 1) - f_2(t)(1 - f_1(t))N\}, \quad (6.2)$$

with $W \equiv \frac{2\pi}{\hbar}M(\frac{\epsilon_1 - \epsilon_2}{\hbar})$.

The important steps in order to obtain this equation were: (i) decoupling of higher order correlations; (ii) equilibrium of the bath (phonons) and (iii) approximation of null memory depth.

Chapter 7

Naive quantum kinetics

The question arises whether it is not better to avoid the last step (iii) in the deduction of the rate equation. The goal is to obtain an equation for the adequate description of rapid processes at short times. Thus we are looking for a pre-Markovian description including memory effects. This means going back from the delta-function

$$\delta(\epsilon_k^0 - \epsilon_{k'}^0 \pm \hbar\omega_q)... \leftrightarrow$$

in the rate equation Eq.(6.2) to

$$\frac{2}{\hbar} \int_{t_0}^{t} dt' \cos(\frac{(t-t')}{\hbar}(\epsilon_k^0 - \epsilon_{k'}^0 \pm \hbar\omega_q))...,$$

and leaving the populations under the time integral. This way, we are led - within the frame of the model from the previous chapter - back to the equation

$$\frac{\partial}{\partial t} f_1(t) = -\frac{2}{\hbar} \int d\omega M(\omega) \int_{t_0}^{t} dt' \cos((\epsilon_2 - \epsilon_1 - \hbar\omega)\frac{(t-t')}{\hbar})$$

$$\times \{f_1(t')(1 - f_2(t'))(\mathcal{N}(\hbar\omega) + 1) - f_2(t')(1 - f_1(t'))\mathcal{N}(\hbar\omega)\}.$$

Yet further modification is needed, since under this simplified form, the equation may not be consistent. In practice, one introduces a supplementary memory depth factor $D(t - t')$, at least in the case of dispersionless phonons (LO-phonons in semiconductors), in order to ensure the asymptotic recovery of the Markovian limit. It is expected, that this factor will appear as an effect of higher order corrections, due to the fact that in the presence of the same interaction, the one-particle states are not stable.

Thus - on intuitive grounds - one introduces the factor

$$D(t - t') \approx e^{-(\gamma_k + \gamma_{k'})(t-t')},$$

containing γ_k -the imaginary part of the energy ϵ_k, or equivalently said, the lifetime ($\frac{\hbar}{\gamma_k}$) of the state k. Numerical simulations of the equation show that this last *ad hoc* ingredient is needed in the case of dispersionless phonons; not only for the recovery of the Markovian behavior for large times, but also to avoid non-physical solutions with the average populations outside the range $(0, 1)$. This may be interpreted as the requirement: the equation should keep track of the occupation of the states only within the lifetime of the states.

With this supplementary ingredient included, we may consider the simplest "quantum-kinetic" equation

$$\frac{\partial}{\partial t} f_1(t) = -\frac{2}{\hbar} \int d\omega M(\omega) \int_{t_0}^t dt' \cos((\epsilon_2 - \epsilon_1 - \hbar\omega)\frac{(t - t')}{\hbar})$$
$$\times D(t - t')\{f_1(t')(1 - f_2(t'))(\mathcal{N}(\hbar\omega) + 1)$$
$$- f_2(t')(1 - f_1(t'))\mathcal{N}(\hbar\omega)\}. \tag{7.1}$$

This equation is of general form

$$\frac{\partial f(t)}{\partial t} = \int_{t_0}^t dt' K(t - t') F(f(t')),$$

where $K(t)$ is the memory kernel and $F(f)$ is a non-linear function of f. Of course, one needs as initial condition $f(t)$ at $t = t_0$. This kind of non-local equation (in the time variable) we have met already by the analysis of the solvable classical models. Again the non-locality is due to fact that we wanted to write down closed equations for some limited degrees of freedom, by eliminating other degrees of freedom. Here the "system" is represented by the population $f(t)$ and the eliminated degrees of freedom are the higher averages like $\langle a_2^\dagger a_1 b_\alpha \rangle$. However, the derivation is in the quantum-mechanical case accompanied by unavoidable rough approximations, like that of decoupling the next higher averages, and ignoring the influence of the interaction on the phonons, as it is expressed in Eq.(6.1).

This type of equation brings some principial new aspects into the time-evolution problem. Suppose we have solved it until a time instant t_1, and would like to use it to find $f(t)$ for later times $t > t_1$ by solving the equation

$$\frac{\partial f(t)}{\partial t} = \int_{t_1}^{t} dt' K(t - t') F(f(t')) \,.$$

Unfortunately it would not coincide with the solution of the original equation, solved at once until this time. Indeed

$$\int_{t_0}^{t} dt' K(t - t') F(f(t')) = \int_{t_0}^{t_1} dt' K(t - t') F(f(t'))$$
$$+ \int_{t_1}^{t} dt' K(t - t') F(f(t')) \,,$$

and we have an additional piece. This new aspect - we already met by the analysis of the solvable classical models - and it may seem at a first glance to be not so relevant, but in the quantum mechanical context it poses the basic conceptual question: which is the correct "initial" time instant? The only possible answer is to extend the memory over the whole prehistory of the system:

$$\frac{\partial f(t)}{\partial t} = \int_{-\infty}^{t} dt' K(t - t') F(f(t')) \,.$$

We shall see in later chapters that in the most important applications of quantum kinetics in the ultra-short-time spectroscopy of semiconductors, the optical pulse sets the clock. The system may be considered as being in the ground state until the arrival of the laser pulse. There is no electron-hole population in the past. The physics of that problem, however, is not exhausted by the carrier population. We have to know the average of the inter-band polarization too. The described quantum kinetics for the populations alone is too naive in this sense.

Concerning the method, the kind of quantum kinetics we described here, is the one based on the equation of motion method, for which it is typical to consider coupled equations for time dependent averages. Instead the consideration of many variables with a single time, one may treat the problem with less variables, but depending on more times. Such entities are the time-correlations, which are the object of the diagrammatic Green function technique, and will be discussed later.

From the very beginning we wanted to give a glimpse at the whole complexity of the problems in describing non-equilibrium. Later we will discuss more detailedly and at a higher level of sophistication most of these aspects.

In the meantime, we shall present the more traditional ways to deal with non-equilibrium problems.

Chapter 8

Diagonal Master equations and rate equations

8.1 Properties of the Master equation

In this chapter, we are going to discuss the properties of the diagonal Master equations and their connection to the rate equations. The probability $P_n(t)$ for a system to be at time t in the eigenstate $|n\rangle$ of the (unperturbed) Hamiltonian H_0 is just the diagonal matrix element of the density operator in this basis. It does not exhaust all the statistical information about the system, but it enables to calculate the averages of all the operators, which commute with H_0. Later we will discuss the generalized Master equations, including also the non-diagonal information. However, the latter belong to more recent developments in the field.

Let $W_{nn'}$ be the transition rates between the states $|n\rangle$ and $|n'\rangle$ due to some perturbation H' (described for example by the "golden rule".) The Master equation is

$$\frac{\partial}{\partial t} P_n(t) = -\sum_{n'} (P_n(t) W_{nn'} - P_{n'}(t) W_{n'n}) \equiv -\sum_{n'} A_{nn'} P_{n'}(t), \quad (8.1)$$

i.e. gain minus loss. This equation describes a so called Markovian process, having a semi-group property. This means that it has no memory effects and the evolution from t_0 to $t_1 > t_0$ can as well be described as an evolution from t_0 to some intermediary time $\bar{t}$ ($t_0 < \bar{t} < t_1$) followed by an evolution from $\bar{t}$ to t_1.

A) Let us consider first the case without thermostat, as it was implicitly admitted in the Gedanken-experiment (see Chapter 5). From the definition of the transition rates Eq.(5.3) it follows that that they are positive

$$W_{nn'} \geq 0,$$

and satisfy the generalized balance relations:

$$\sum_{n'} (W_{nn'} - W_{n'n}) = 0.\tag{8.2}$$

This general balance property results from the unitarity of the time evolution operator U. A more strict relationship $W_{nn'} = W_{n'n}$ occurs however in the lowest order perturbation theory. The above properties dictate also the basic features of the Master equation and of its solutions:

1) *Conservation of the normalization:*

$$\frac{\partial}{\partial t} \sum_{n'} P_{n'}(t) = 0.$$

Therefore, one may take $\sum_{n'} P_{n'}(t) = 1$ for all times.

2) *Conservation of positivity:* If $P_n(0) \geq 0$, then it is always positive, $P_n(t) \geq 0$ for any $t > 0$. This results from the fact that $P_n(t) = 0$ implies $\frac{\partial}{\partial t} P_n(t) \geq 0$.

3) *Unique asymptotic equilibrium.* The formal solution of Eq.(8.1) is

$$P(t) = e^{-At} P(0).$$

Now, the real part of the eigenvalues of A are non-negative. Indeed, with $\xi_n = x_n + \imath y_n$ a complex number one has

$$\Re \left\{ \sum_{n,n'} \xi_n^* A_{nn'} \xi_{n'} \right\} = \Re \left\{ \sum_{n,n'} W_{nn'} \xi_n^* (\xi_n - \xi_{n'}) \right\}$$

$$= \frac{1}{2} \sum_{n,n'} W_{n'n} \left[(x_n - x_{n'})^2 + (y_n - y_{n'})^2 \right] \geq 0,$$

where we used the generalized balance Eq.(8.2) from which the non-negativity of the eigenvalues follow. If a general connectivity holds, i.e. every state may be reached from every state by a chain of transitions, then the above inequality implies even that one has a unique null-eigenvector, and this has equal components $\xi_n^0 = const$. It follows then

$$\lim_{t \to \infty} P(t) = P^0,$$

with the asymptotic equilibrium probability P^0 having equal components P^0_n. We can also write

$$P(t) = e^{-At}(P(0) - P^0) + P^0 \,,$$

indicating the decay of the non-equilibrium part.

B) Now let us split the whole system into the system proper and a thermal bath. We shall denote the states of the system proper as before with the index n, while α denotes a state of the thermostat. Let us suppose that the evolution of the probability $P_{n\alpha}(t)$ for whole system is described again by a Master equation like Eq.(8.1), but with the indices n, α. Under the rough assumption that the thermal bath is always in equilibrium

$$P_{n\alpha}(t) = \mathcal{P}_n(t)\mathcal{P}^0_\alpha; \quad \mathcal{P}^0_\alpha = \frac{e^{-\beta E_\alpha}}{\sum_{\alpha'} e^{-\beta E_{\alpha'}}} \,, \tag{8.3}$$

one may write a closed Master equation for the system proper. Taking a summation over the thermostat index α it follows for $P_{n\alpha}(t)$ the equation

$$\frac{\partial}{\partial t}\mathcal{P}(t) = -\mathcal{A}\mathcal{P}(t); \quad \mathcal{A}_{nn'} = \delta_{nn'}\sum_{n''}\mathcal{W}_{nn'} - \mathcal{W}_{n'n} \,, \tag{8.4}$$

with the averaged transition rates

$$\mathcal{W}_{nn'} \equiv \sum_{\alpha\alpha'} \mathcal{P}^0_\alpha W_{n\alpha,n'\alpha'} \,. \tag{8.5}$$

If we assume the conservation of the total unperturbed energy ($E_n + E_\alpha = E_{n'} + E_{\alpha'}$) as it is given by the "golden rule" and the generalized balance Eq.(8.2) (written in the total index n, α) the generalized balance with thermostat follows

$$\sum_{n'}\left(\mathcal{W}_{nn'} - e^{\beta(E_n - E_{n'})}\mathcal{W}_{n'n}\right) = 0 \,. \tag{8.6}$$

The operator $\mathcal{A}$ has again only eigenvalues with a non-negative real part and under similar connectivity conditions the only null-eigenvector (in the space with a fixed number of particles) is

$$P^0_n = \frac{e^{-\beta E_n}}{\sum_{n'} e^{-\beta E_{n'}}} \,.$$

Actually we got the new Master equation for the system in contact with the thermostat, without checking the validity of the factorization Eq.(8.3). This remains itself the definition of the thermal bath. This assumption is plausible only if the thermal bath is infinite, and the system is finite. The whole description of the Master equation, as it is given here - as coherent as it seems - has nothing to do with a mathematical derivation. Let us not forget that the assumed properties of the transition rates are the ones resulting from a Gedanken-experiment. The Master equation is, however, a consistent description of evolution based on physical intuition, and was often used successfully, although its mathematical derivations came relatively late. We shall discuss later these recent derivations too, but for the moment we are going to concentrate on the connection between the Master equation and the rate equations for the average occupation of one-particle states.

8.2   Rate equations

Let us consider now the case, where the index n describes the occupation numbers (for fermions) $n_i = 0, 1$ of the one-particle states $|i\rangle$

$$n \equiv \{n_1, n_2, \ldots\}.$$

The energy (without interaction) is

$$E_n = \sum_i e_i n_i.$$

The transition rates due the interaction with a thermal bath (for example phonons) in the lowest order perturbation theory ("golden rule") are a sum of one-particle processes

$$W_{nn'} = \sum_{i,j} w_{ij} \delta_{n_i, n'_i+1} \delta_{n'_j, n_j+1} \Pi_{l \neq i,j} \delta_{n_l, n'_l}, \qquad (8.7)$$

and a "detailed balance" relation $w_{ij} = w_{ji} e^{\beta(e_i - e_j)}$ holds.

We want to derive, starting from the Master equation, the equation for the average occupation numbers

$$\bar{n}_i(t) \equiv \sum_n n_i \mathcal{P}_n(t) \,.$$

From the Master equation it follows immediately

$$\frac{\partial}{\partial t} \bar{n}_i(t) = -\sum_{n,n'} W_{nn'}(n_i - n_i') \mathcal{P}_n(t) = -\sum_{i,j} \left[w_{ij} \overline{n_i(1-n_j)} - w_{ji} \overline{n_j(1-n_i)} \right] \,,$$

where the bars are indicating the averaging, i.e.

$$\overline{n_i(1-n_j)} \equiv \sum_n \mathcal{P}_n n_i(1-n_j) \,.$$

This gives rise to a closed set of equations only after the factorization approximation

$$\overline{n_i n_j} \approx \bar{n}_i \bar{n}_j \,,$$

which neglects correlations.

Thus for fermions we get the rate equation

$$\frac{\partial}{\partial t} \bar{n}_i(t) = -\sum_{i,j} \left[w_{ij} \bar{n}_i(t)(1 - \bar{n}_j(t)) - w_{ji} \bar{n}_j(t)(1 - \bar{n}_i(t)) \right] \,. \tag{8.8}$$

As we can see, the Pauli exclusion principle is taken into account only on the average. However, the equation still keeps the average occupation numbers in the interval $[0, 1]$, and - under the assumption of connectivity of the transition rates - leads asymptotically to the Fermi-function, the only stable fixed point (and stationary solution) of the equations.

In the case of bosons a similar equation may be derived, but the $1 - \bar{n}_j(t)$ factors for the final states are then replaced by an $1 + \bar{n}_j(t)$ factor, and the asymptotic solution is then the Bose distribution.

Let us now linearize the equation in the neighborhood of the equilibrium solution, admitting that the initial state is not far away from equilibrium. Then, we look for a small deviation $\delta\bar{n}_i(t)$ from the equilibrium Fermi-function $f_i \equiv f(\epsilon_i)$

$$\delta\bar{n}_i(t) \equiv \bar{n}_i(t) - f_i \,.$$

Keeping only the terms that are linear in this deviation, one has

$$\frac{\partial}{\partial t}\delta\bar{n}(t) = -\Gamma\delta\bar{n}(t)\,,$$

with the matrix notation

$$\Gamma_{ij} \equiv \left(\delta_{ij}\sum_{l}\mathcal{W}_{il} - \mathcal{W}_{ij}\right)\frac{1}{f_i(1-f_i)}\,,$$

and

$$\mathcal{W}_{ij} = f_i(1-f_j)w_{ij} = \mathcal{W}_{ji}\,.$$

The matrix Γ is real and hermitian in the scalar product

$$(\psi,\phi) \equiv \sum_{i}\psi_i^*\phi_i\frac{1}{f_i(1-f_i)}\,,$$

and it is also non-negative since

$$(\psi,\Gamma\psi) = \frac{1}{2}\sum_{i,j}\left|\frac{\psi_i}{f_i(1-f_i)} - \frac{\psi_j}{f_j(1-f_j)}\right|^2\mathcal{W}_{ij} \geq 0\,.$$

Again, under the assumption of connectivity one has only a single null-eigenvector

$$\xi_i = \frac{f_i(1-f_i)}{\sum_j\sqrt{f_j(1-f_j)}}\,.$$

Therefore, the deviation

$$\delta\bar{n}(t) = e^{-\Gamma t}\delta\bar{n}(0)$$

always disappears asymptotically ($t \to \infty$ $\delta\bar{n}(t) \to 0$). The only null-eigenvector corresponds to the particle number conservation

$$\sum_{i}\delta\bar{n}(t) = 0\,,$$

which is equivalent to

$$(\xi, \delta\bar{n}(t)) = 0\,,$$

i.e. $\delta\bar{n}(t)$ has no components along the direction of the vector ξ having constant components.

We are going to emphasize that the calculation of the occupation number time-correlation function, given by the Master equation, defined as

$$\phi_{ij}^M(t) \equiv \sum_{n,n'} P_n^0 \left(e^{-At}\right)_{n'n} n_i n_j'\,,$$

may be approximatively reduced also to a rate equation problem [Bányai and Aldea (1979)]. Indeed, it can be written in the form

$$\phi_{ij}^M(t) = \frac{d}{d\lambda}\bar{n}_j^\lambda(t)|_{\lambda=0} + f_i f_j\,,$$

where

$$\bar{n}_j^\lambda(t) = \sum_n n_j \mathcal{P}_n^\lambda(t)\,,$$

$\mathcal{P}_n^\lambda(t)$ being the solution of the Master equation Eq. (8.4) with the initial condition

$$P_n^\lambda(0) = P_n^0 \left[1 + \lambda(n_i - f_i)\right]\,,$$

satisfying $\sum_n P_n^\lambda(0) = 1$, and $P_n^\lambda(0) \geq 0$ for $\lambda \leq 1$. In the linearized rate equation approximation for $\bar{n}_j^\lambda(t)$, with the corresponding initial condition

$$\bar{n}_j^\lambda(0) = f_j + \lambda\delta_{ij}f_j(1 - f_j)\,,$$

one gets then the Master-Boltzmann approximation

$$\phi_{ij}^M(t) \approx \phi_{ij}^{MB}(t) = f_i(1 - f_j)\left(e^{-\Gamma t}\right)_{ij} + f_i f_j\,. \tag{8.9}$$

This formula has important applications in the theory of the electric conductivity we discuss in later chapters.

8.3　Scattered versus hopping electrons

Let us give here two important examples for the use of rate equations. The first one is the known case of free electrons scattered due to interaction with a thermal bath of phonons. In this case the index i (quantum number) of the previous general discussion characterizing the unperturbed electron, is the wave-vector $\vec{k}$. The energy of this state is $\epsilon_{\vec{k}} = \frac{\hbar^2 k^2}{2m}$. The knowledge of the average occupation of the state $\vec{k}$ allows the calculation of the average of any dynamical variable, which is diagonal in $\vec{k}$, such as for example the averaged homogenous current

$$\langle \vec{j} \rangle_t = \frac{e}{V} \sum_{\vec{k}} \vec{k} \bar{n}_{\vec{k}}(t) \,,$$

and the average energy density (per volume V)

$$\langle E \rangle_t = \frac{1}{V} \sum_{\vec{k}} \frac{\hbar^2 k^2}{2m} \bar{n}_{\vec{k}}(t) \,.$$

The second important model is the dual model of localized electrons on some sites of positions $\vec{x}_i$. (Here the index i shows only that these sites are discrete.) The energies of the electrons on these sites are ϵ_i . When localization occurs due to disorder, both the positions and energies are randomly distributed. These electrons are hopping form site-to site due to absorption or emission of a phonon. In this case, the rate equation allows to calculate the average of any entity that depends only on the positions of the electrons, like the motion of the average center of mass

$$\langle \vec{X} \rangle_t = \frac{1}{N} \sum_{i=1}^{N} \vec{x}_i \bar{n}_i(t) \,,$$

and the average energy (per number N of particles)

$$\langle E \rangle_t = \frac{1}{N} \sum_{i=1}^{N} \epsilon_i \bar{n}_i(t) \,.$$

These two extreme models are successfully used in the theory of transport phenomena as we shall see it in later chapters.

Chapter 9

The Boltzmann equation

The rate-equations (or the Master equation itself) do not allow the calculation of averages of operators, which do not commute with the unperturbed Hamiltonian of the system. For example, in the case of electrons weakly interacting with phonons, one may compute with the rate equation the average occupations of the $\vec{k}$ -states, therefore also the average overall homogeneous flow, but not local charge and current densities in the case of an inhomogeneous flow. The situation is essentially different in the classical statistical mechanics. Here helps the well-known Boltzmann equation.

Within the frame of classical statistical mechanics, one considers the probability density $p(\vec{v}, \vec{x}, t)$ to find a particle having the velocity $\vec{v}$ and coordinate $\vec{x}$ at time t. This probability density is normalized (in the volume V)

$$\int d\vec{v} \int_V d\vec{x} \, p(\vec{v}, \vec{x}, t) = 1 \, .$$

From the well-known classical Liouville equation for charged particles moving in the electric $\vec{E}(\vec{x}, t)$ and magnetic $\vec{B}(\vec{x}, t)$ fields it follows

$$\left(\frac{\partial}{\partial t} + \vec{v} \frac{\partial}{\partial \vec{x}} + \frac{1}{m} \vec{F}(\vec{x}, t) \frac{\partial}{\partial \vec{v}} \right) p(\vec{k}, \vec{x}, t) = 0 \, ,$$

where $\vec{F} = e(\vec{E} + \vec{v} \times \vec{B})$ is the Lorenz force. This equation expresses the fact that along the path of the particle the probability remains unchanged.

In the following, we consider only an electric field. Since the N charged particles in the volume V are also sources of a field, one needs the

Poisson equation to define the total field in a self-consistent manner

$$\epsilon_0 \frac{\partial}{\partial \vec{x}} \vec{E}(\vec{x}, t) = 4\pi e N \left(\int d\vec{v} p(\vec{v}, \vec{x}, t) - \frac{1}{V} \right) + 4\pi \rho_{ext}(\vec{x}, t) . \qquad (9.1)$$

Besides the internal charge density (with a compensating homogeneously distributed background charge)

$$\rho(\vec{x}, t) \equiv e N \left(\int d\vec{v} p(\vec{v}, \vec{x}, t) - \frac{1}{V} \right) ,$$

we consider here also an external inhomogeneous charge density $\rho_{ext}(\vec{x}, t)$.

Now we must take into account also that due to interaction with a thermal bath (phonons) or to elastic scattering on some disordered potential a stochastic collision term has to be introduced. We put this collision term on the right-hand side of the equation, analogously to the case of a pure rate (or Master) equation. It looks as if the collisions would take place homogeneously in space

$$\left(\frac{\partial}{\partial t} + \vec{v} \frac{\partial}{\partial \vec{x}} + \frac{e}{m} \vec{E}(\vec{x}, t) \frac{\partial}{\partial \vec{v}} \right) p(\vec{v}, \vec{x}, t) = \frac{\partial p(\vec{v}, \vec{x}, t)}{\partial t} \Big|_{coll} \qquad (9.2)$$

$$\frac{\partial p(\vec{v}, \vec{x}, t)}{\partial t} \Big|_{coll} = - \int d\vec{v}' \left[w(\vec{v}, \vec{v}') p(\vec{v}, \vec{x}, t) - w(\vec{v}', \vec{v}) p(\vec{v}', \vec{x}, t) \right] .$$

In the case of elastic collisions the transition rates conserve the kinetic energy

$$w(\vec{v}, \vec{v}') = w(\vec{v}', \vec{v}) \quad (e(\vec{v}) = e(\vec{v}')) , \qquad (9.3)$$

while in the case of phonon-emission or absorption the detailed balance

$$w(\vec{v}, \vec{v}') = e^{\beta(e(\vec{v}) - e(\vec{v}'))} w(\vec{v}', \vec{v}) \qquad (9.4)$$

holds, with $e(\vec{v}) \equiv \frac{mv^2}{2}$ being the kinetic energy of the particles, and β is the inverse temperature of the phonons.

This equation may be obtained (however, not rigorously) from the quantum-mechanical frame by a development of the Wigner-function in powers of $\hbar$. The transition rates are then defined by the "golden rule".

The above equations conserve the average total number of particles

$$\frac{\partial}{\partial t} \int d\vec{v}d\vec{x}p(\vec{v}, \vec{x}, t) = 0 \,,$$

and in the case of the elastic collisions with time independent external charge density ρ_{ext} also the average total energy:

$$\frac{\partial}{\partial t} \left(\frac{N}{V} \int d\vec{v}d\vec{x}p(\vec{v}, \vec{x}, t)(e(\vec{v}) + U_{ext}(\vec{x})) + \frac{1}{2} \int d\vec{x} \int d\vec{x}' \frac{\rho(\vec{x}, t)\rho(\vec{x}', t)}{\epsilon_0 |\vec{x} - \vec{x}'|} \right) = 0 \,,$$

or equivalently

$$\frac{\partial}{\partial t} \left(\frac{N}{V} \int d\vec{v}d\vec{x}p(\vec{v}, \vec{x}, t) \left(e(\vec{v}) + U(\vec{x}, t)\right) - \frac{1}{2} \int d\vec{x} \int d\vec{x}' \frac{\rho(\vec{x}, t)\rho(\vec{x}', t)}{\epsilon_0 |\vec{x} - \vec{x}'|} \right) = 0,$$

where

$$\vec{E}(\vec{x}, t) \equiv -\nabla U(\vec{x}, t) \,,$$

and

$$\rho(\vec{x}, t) \equiv eN \left(\int d\vec{v}p(\vec{v}, \vec{x}, t) - \frac{1}{V} \right) \,.$$

We can see here also the role of the Hartree-constant, like in the quantum-mechanical theory.

The collisions between the particles themselves may be described by a more complicated structure

$$\frac{\partial p(\vec{v}, \vec{x}, t)}{\partial t}\Big|_{coll} = - \int d\vec{v}' \int d\vec{v}_1 \int d\vec{v}_1' \, [W(\vec{v}, \vec{v}'; \vec{v}_1, \vec{v}_1')p(\vec{v}, \vec{x}, t)p(\vec{v}', \vec{x}, t)$$
$$- W(\vec{v}_1, \vec{v}_1'; \vec{v}, \vec{v}')p(\vec{v}_1, \vec{x}, t)p(\vec{v}', \vec{x}, t)] \,.$$

Here, besides the kinetic energy the total momentum is also conserved.

In a finite volume with vanishing boundary conditions, in a constant external field we get an asymptotic equilibrium state. This results from the conservation laws and the properties of the transition rates (symmetry, "detailed balance" and connectivity).

In the case of elastic collisions the micro-canonical distribution

$$p_0(\vec{v}, \vec{x}, t) = const. \qquad for \qquad \frac{mv^2}{2} + U(\vec{x}) = const.$$

results, while in the case of inelastic scattering on phonons the Maxwell-Boltzmann distribution

$$p_0(\vec{v}, \vec{x}, t) = \frac{1}{VZ} e^{-\beta(\frac{mv^2}{2} + U(\vec{x}))}$$

emerges.

It is worth noticing that in both cases $U(\vec{x})$ is understood as the asymptotic self-consistent potential energy (in equilibrium).

In an infinite volume, or with periodic boundary conditions, one obtains in the presence of a d.c. field a stationary flow state.

Actually, the Boltzmann theory, as it is used in condensed matter theory, contains elements of quantum mechanics, since the transition rates are calculated with the help of the quantum mechanical "golden rule". Furthermore, we need another quantum mechanical ingredient related to particle statistics. In order to do this, one defines

$$f(\vec{p}, \vec{x}, t) = \langle N \rangle \frac{1}{(2\pi\hbar m)^3} p(\vec{v}, \vec{x}) \,,$$

which is normalized to the total average number of particles $\langle N \rangle$ as

$$\frac{1}{(2\pi\hbar)^3} \int d\vec{x} \int d\vec{p} f(\vec{p}, \vec{x}, t) = \langle N \rangle \,.$$

Obviously, $f(\vec{p}, \vec{x}, t)$ represents the average number of particles in the "elementary cell" $(2\pi\hbar)^3$ of the phase space around the phase-space point $\vec{x}, \vec{p}$. Thereafter one introduces in the collision term final state factors by replacing $f(\vec{p}, \vec{x}, t)$ by $f(\vec{p}, \vec{x}, t)(1 \pm f(p', \vec{x}, t))$ and $f(\vec{p}', \vec{x}, t)$ by $f(\vec{p}', \vec{x}, t)(1 \pm f(\vec{p}, \vec{x}, t))$ depending on the Bose or Fermi Statistics. Such equations may be motivated again by identifying $f(\vec{p}, \vec{x}, t)$ within quantum mechanics with the Wigner function

$$f(\vec{p}, \vec{x}) = \int d\vec{y} e^{\frac{i\vec{p}\vec{y}}{\hbar}} \langle \psi(\vec{x} + \frac{1}{2}\vec{y}, t)^+ \psi(\vec{x} - \frac{1}{2}\vec{y}, t) \rangle \,,$$

(where $\psi(\vec{x}, t)$ is the second quantized wave function and $\langle \ldots \rangle$ means averaging over a given ensemble), and implementing a quasi-classical approximation.

Chapter 10

Hydrodynamics as scaling of kinetic theory

The ultimate goal of statistical mechanics is to deduce the properties of macroscopical bodies from the underlying microscopical dynamics of their constituents. However, this program - as we already mentioned - encounters a formidable obstacle, related to the contradiction between the microscopical reversibility and macroscopical irreversibility. In order to facilitate the task, an intermediate level of description is introduced (the kinetic level) that operates with microscopical entities, but within an irreversible description. To this category belong the already discussed Boltzmann equation, different Master and rate equations, as well as the Langevin equations etc.

Our purpose now is to discuss the passage from the kinetic to the macroscopic (hydrodynamic) level. Its importance lies not only in fundamental understanding, but also in providing clues to the correct calculation of phenomenological constants as well.

Within the frame of the Boltzmann theory of neutral gases, already Hilbert was preoccupied by this problem, while Chapman and Enskog gave an expansion method that proved to be very fruitful. Their basic idea was that after a certain time, the state approaches a local equilibrium, in which the distribution function depends on time only through the densities of the conserved entities.

Whilst working out the connection between his microscopical electrodynamics and Maxwell's macroscopical one, Lorentz felt the need for some smoothening of the microscopical entities in space. While the microscopical field varies rapidly on the interatomic distance, the corresponding macroscopical field varies only on a macroscopical scale.

These two basic ideas of "smoothening" in time and space are implemented in the scaling theories of the connection between the kinetic and hydrodynamic descriptions. The key of the fundamental understanding of the problem seems to be "looking at" the microscopical solutions on the macroscopical scale of length and time. The mathematical formulation was suggested first by the scale invariance of the solution of the diffusion equation, and then by the more complicated scale invariance of the solution of the diffusion-conduction equation describing the charge flow in a semiconductor. These ideas have been rigorously implemented so far only in the discussion of linear hydrodynamics (i.e. small departure from equilibrium) of some exactly solvable kinetic models (neutral Lorentz gas, charged Boltzmann gas with instant thermalization and charged hopping lattice gas) [Bányai (1984)]. Nevertheless, these exact results on essentially different models give a hint to the basic mechanisms as well as to the generality of the approach.

In order to get a clear idea of the problem we are facing, let us consider the diffusion of neutral particles in a medium. Macroscopically it is described by the diffusion equation

$$\frac{\partial}{\partial t} n = D\nabla^2 n \,,$$

for the particle density $n(\vec{x}, t)$, while microscopically (at the kinetic level) we may have a Boltzmann model for the distribution function $p(\vec{v}, \vec{x}, t)$ of the velocities and coordinates

$$\left(\frac{\partial}{\partial t} + \vec{v} \frac{\partial}{\partial \vec{x}} \right) p(\vec{v}, \vec{x}, t) = \frac{\partial p(\vec{v}, \vec{x}, t)}{\partial t} \Big|_{coll} \,,$$

with the collision term describing scattering on thermal scatterer. The particle density in this case is given by integrating over the velocities, and multiplying with the total average number of the particles

$$n^{micro}(\vec{x}, t) = \langle N \rangle \int d\vec{v} p(\vec{v}, \vec{x}, t).$$

As one can see immediately, in the microscopic (kinetic) model the particle density at a given instant is determined by giving the full coordinate-velocity distribution at some initial instant $t = 0$. In the macroscopical (hydrodynamical) description, however, one renounces to know everything about the system; yet the entity thus retained $n(\vec{x}, t)$ depends only on its own initial value. Therefore, we want less information, under the condition that the same amount of information can be discarded at $t = 0$. It is clear that in order to implement this idea, some operation has to be performed on the microscopical density. We start from the observation that the diffusion equation has a certain scale invariance: its basic solution (corresponding to $n(\vec{x}, 0) = \langle N \rangle \delta(\vec{x})$) is invariant against the simultaneous scaling of the coordinate $\vec{x} \to \lambda \vec{x}$ and time $t \to \lambda^2 t$, and a subsequent multiplication with λ^3. On the other hand, the operation

$$\lim_{\lambda \to \infty} \lambda^3 n^{micro}(\lambda \vec{x}, \lambda^2 t)$$

produces a function that is invariant against further finite scalings (it is a scale projector) and as we shall show it, performs also the task we are looking for. Then, one may expect that

$$\lim_{\lambda \to \infty} \lambda^3 n^{micro}(\lambda \vec{x}, \lambda^2 t) = n(\vec{x}, t) \,,$$

or reformulating the statement

$$\lim_{\lambda \to \infty} \lambda^3 \left[n^{micro}(\lambda \vec{x}, \lambda^2 t) - n(\lambda \vec{x}, \lambda^2 t) \right] = 0 \,,$$

i.e. the difference between the microscopical and macroscopical densities at large distances $\lambda \vec{x}$ and large times $\lambda^2 t$ goes to zero faster than $\frac{1}{\lambda^3}$.

We shall see later that - generally speaking - one has to scale all the macroscopical lengths including the Debye length in a semiconductor, which is of electromagnetic nature. Such a scaling accomplishes also the transition to a continuum description for the lattice models we treat.

In the first Section we will discuss the macroscopical electrodynamic description of the charge flow in a semiconductor. Then we introduce the scale projection conjecture and the general mathematical features of its implementation, anticipating some common features of the microscopical (kinetic) models we analyze later more detailedly. The third Section is devoted to the thorough analysis of a neutral hopping model on a periodical lattice. The fourth Section contains a presentation of the results on the charged hopping model, while the last Section contains a sketchy presentation of the proofs in the case of a charged Boltzmann gas with "instant thermalizing" cross sections.

10.1   Macroscopic description of charge evolution in a semi-conductor

Let us consider an infinite homogeneous semiconductor medium, characterized by the phenomenological constants: σ(conductivity), D (diffusivity) and ϵ (dielectric permittivity). We do not require isotropy, and therefore all these entities are actually 3×3 tensors. No external magnetic field is introduced, and the internal magnetic field will be ignored (small internal currents). The relevant Maxwell equations are:

the continuity equation

$$\frac{\partial \rho_T}{\partial t} + \nabla \vec{j}_T = 0 \,, \tag{10.1}$$

for the total charge and current densities and

the Poisson equation

$$\nabla \vec{E} = 4\pi \rho_T \,, \tag{10.2}$$

relating the electric field to the total charge density.

The total charge and current densities are made up of external and internal ones

$$\rho_T = \rho_{int} + \rho_{ext}$$
$$\vec{j}_T = \vec{j}_{int} + \vec{j}_{ext} \,.$$

We chose to have no external magnetic fields, therefore $\vec{j}_{ext} = 0$ and $\frac{\partial \rho_{ext}}{\partial t} = 0$, i.e. the external charge we consider is static. This implies also that the electric field stems from a scalar potential $\vec{E} = -\nabla V$.

Now, the phenomenological relations specific for a semiconductor may be written as

$$\rho_{int} = \rho - \frac{1}{4\pi} \nabla (\epsilon - 1) \vec{E} \,, \tag{10.3}$$

$$\vec{j}_{int} = \sigma \vec{E} - D \nabla \rho + \frac{1}{4\pi} (\epsilon - 1) \frac{\partial \vec{E}}{\partial t} \,. \tag{10.4}$$

Here ρ represents the so-called "free" charge density. We will see that it needs no definition. Once the initial "free" charge density ρ is given, these phenomenological equations together with the above Maxwell equations completely determine all the physical entities, everywhere and at any instant.

Before writing down the combined equations for the two basic entities ρ and V, it is useful to remember that there is a universal relationship (the Einstein relation) known between the conductivity and diffusivity

$$\sigma = e^2 \frac{\partial n_0}{\partial \mu} D \,,$$

or

$$\sigma = \frac{1}{4\pi} \ell^{-2} D \,.$$

Here we introduced the electromagnetic length ℓ as given by

$$\ell = \left(4\pi e^2 \frac{\partial n_0}{\partial \mu} \right)^{-\frac{1}{2}} \,,$$

where e, n_0 are the charge and the equilibrium density of the "free" carriers, while μ is their chemical potential. This universal relation stems from the fact that equilibrium may subsist in the presence of an external charge due to the compensation of the conduction current due to the diffusion current. This relationship will be discussed later within the frame of the linear response theory.

Then, introducing Eqs.(10.3)-(10.4) into Eqs. (10.1) and (10.2) we get

$$\frac{\partial \rho}{\partial t} = \nabla D \nabla (\rho + \frac{1}{4\pi\ell^2} V) \,, \tag{10.5}$$

$$\nabla \epsilon \nabla V = -4\pi \left(\rho + \rho_{ext} \right) \,, \tag{10.6}$$

which confirm our statement that the only initial condition, which has to be given is $\rho|_{t=0}$. On the other hand, giving $\rho|_{t=0}$ is equivalent to giving $\rho_{int}|_{t=0}$. Therefore, there is no need for any special definition of the "free" charge density. In the phenomenological theory it gives just a convenient way to describe the properties of the material. However, as it was anticipated by Lorentz, and as we will see on some examples, in a microscopical theory it has a well-defined meaning, corresponding to the charge that can be transported over macroscopical distances (in opposition to the polarization charge density).

In an isotropic medium both the diffusion tensor and the dielectric permittivity are diagonal, and one gets a simple closed equation for the "free" charge density

$$\frac{\partial \rho}{\partial t} = D\nabla^2 \rho - \frac{D}{\ell_D^2}(\rho + \rho_{ext}) \, ,$$

where

$$\ell_D = \epsilon^{\frac{1}{2}} \ell$$

is the Debye length, characterizing the space variation of $\rho(\vec{x}, t)$. It obviously coincides with the penetration depth of a static electric field into the semiconductor. It is well-known that this penetration depth is of the order of centimeters in a typical semiconductor, i.e. it is a macroscopical length. This is due to the small concentration of "free" carriers in a semiconductor.

On the contrary, in a metal, due the high concentration of "free" carriers, l_D is microscopical (Angströms). Therefore, phenomena in a metal occur on the surface and not in the bulk. The correct description of these phenomena, however, is not given by our equations. (Macroscopical variations on a microscopical scale just show the breakdown of the description.)

The solution of Eqs.(10.5)-(10.6) is immediate in Fourier transform in the space variable, even in the anisotropic case. We give here the expression of the total charge density

$$\tilde{\rho}_T(\vec{k}, t) = \frac{k^2}{\vec{k}\epsilon\vec{k}}\left[\tilde{\rho}(\vec{k}, 0) + \frac{1}{1 + \ell^2\vec{k}\epsilon\vec{k}}\tilde{\rho}_{ext}(\vec{k})\right] e^{-\vec{k}D\vec{k}(1+\frac{1}{\ell^2\vec{k}\epsilon\vec{k}})t}$$
$$+ \frac{\ell^2 k^2}{1 + \ell^2\vec{k}\epsilon\vec{k}}\tilde{\rho}_{ext}(\vec{k}) \, . \tag{10.7}$$

We would like to call the attention to two remarkable properties:

(i) The polarization charge is instantaneous ($\tilde{\rho}_T(\vec{k}, 0) \neq \tilde{\rho}(\vec{k}, 0) + \tilde{\rho}_{ext}(\vec{k})$).

(ii) The external charge is screened in equilibrium ($\rho_T(0, \infty) = 0$). Nonconservation of charge occurs at the expense of surface charges at infinity.

The fundamental solution corresponds to choosing an initial "free" point charge e at the origin and an external point charge at a position $\vec{x}_0$:

$$\rho(\vec{x}, 0) = e\delta(\vec{x}); \quad \rho_{ext}(\vec{x}) = q\delta(\vec{x} - \vec{x}_0) \, ,$$

or

$$\tilde{\rho}(\vec{k}, 0) = e; \quad \tilde{\rho}_{ext}(\vec{k}) = q e^{i\vec{k}\vec{x}_0} .$$

It is easy to see the scale invariance of the total charge density corresponding to the fundamental solution

$$\tilde{\rho}_T(\frac{1}{\lambda}\vec{k}, \lambda^2 t; \lambda\ell, \lambda\vec{x}_0) = \tilde{\rho}_T(\vec{k}, t; \ell, \vec{x}_0) .$$

Similarly, one may see that

$$\lambda^{-2}\tilde{V}(\frac{1}{\lambda}\vec{k}, \lambda^2 t; \lambda\ell, \lambda\vec{x}_0) = \tilde{V}(\vec{k}, t; \ell, \vec{x}_0) .$$

In the coordinate space these correspond to the scale invariance:

$$\lambda^3 \rho_T(\lambda\vec{x}, \lambda^2 t; \lambda\ell, \lambda\vec{x}_0) = \rho_T(\vec{x}, t; \ell, \vec{x}_0) ,$$
$$\lambda V(\lambda\vec{x}, \lambda^2 t; \lambda\ell, \lambda\vec{x}_0) = V(\vec{x}, t; \ell, \vec{x}_0) .$$

Therefore, all the macroscopical physical entities are invariant with respect to the simultaneous scaling of all lengths (including ℓ) with λ and of the time with λ^2, up to an overall power of λ. Formulated in this way, the statement is valid for an arbitrary solution of our equations.

The reader may remark that the way we treated the macroscopical problem was chosen to be as close as possible to the corresponding microscopical formulation. A peculiar aspect is that one can reconstruct all the physical entities $(\rho_T, \vec{j}_T, \vec{E})$ from the knowledge of the "free" charge density. This is particularly important, since generally speaking, in the microscopical theory the current density might not be uniquely defined. Our discussion has shown that we may restrict our attention to the charge densities.

10.2 The scale projection conjecture

Since the solution of the microscopical (kinetic) problem is not scale invariant, we can only expect that its scale-invariant part may coincide with the solution of the macroscopical problem. The scale-invariant part is obtained by the scale projection

$$\lim_{\lambda \to \infty} \lambda^2 \rho_T^{micro}(\lambda\vec{x}, \lambda^2 t; \lambda\ell, \lambda\vec{x}_0) .$$

If this projection exists, the resulting function is invariant with respect to further scaling with a finite λ.

Though the necessity of scale projection is obvious, it is less obvious that it is sufficient to get the macroscopical result. Our conjecture is that it is sufficient, i.e.

$$\lim_{\lambda\to\infty} \lambda^2 \rho_T^{micro}(\lambda\vec{x}, \lambda^2 t; \lambda\ell, \lambda\vec{x}_0) = \rho_T(\vec{x}, t; \ell, \vec{x}_0).$$

Then we need to identify also the "free" carriers in the microscopical theory to be able to define ℓ, but the important aspect is that once this choice has been made, the calculation of ℓ is immediate.

The conjecture may be also formulated as

$$\lim_{\lambda\to\infty} \lambda^2 \left[\rho_T^{micro}(\lambda\vec{x}, \lambda^2 t; \lambda\ell, \lambda\vec{x}_0) - \rho_T(\lambda\vec{x}, \lambda^2 t; \lambda\ell, \lambda\vec{x}_0) \right] = 0.$$

Under this form, we can see better the physical meaning of the scaling. The above equation tells us that on the big (macroscopical) length and time scales (as λ respectively λ^2) , the two solutions converge (faster than λ^{-3}). On this scale one does not observe the microscopical variations, hence the redundant information is lost. The fact that the characteristic length ℓ should be also scaled, corresponds to the fact that a semiconductor is defined by a macroscopical Debye length. This conjecture is in perfect agreement with the ideas of Lorentz and Chapman-Enskog.

The scaling conjecture has been successfully tested on two different kinds of microscopical (kinetic), exactly solvable models: a) Boltzmann gas of charged particles with "instantaneous thermalization" scattering cross-section, and b) charged hopping particles on a periodical (non-Bravais) lattice. In both cases Coulomb interactions are treated in the self-consistent potential approximation.

These solvable models - which we will analyze in the following part - have some common features, which may be summarized as follows:

The state of the system is statistically characterized by some entity $\mathcal{P}(t)$ satisfying (not far from equilibrium) a linear homogeneous equation

$$\frac{d\mathcal{P}(t)}{dt} = -\mathcal{A}\mathcal{P} + \mathcal{B}\mathcal{V}^{ext}, \tag{10.8}$$

the non-homogeneity being linear in the external potential $\mathcal{V}^{ext}$. The internal charge density results through a linear operation on $\mathcal{P}(t)$:

$$\rho_{int}^{micro}(t) = \mathcal{H}\mathcal{P}(t)\,.$$

The formal solution of Eq.(10.8) may be written as

$$\mathcal{P}(t) = e^{-\mathcal{A}t}\left(\mathcal{P}(0) - \mathcal{P}(\infty)\right) + \mathcal{P}(\infty)\,,$$

and the existence of a unique equilibrium solution $\mathcal{P}(\infty) \equiv \mathcal{A}^{-1}\mathcal{B}\mathcal{V}^{ext}$ results from the assumed irreversibility of the model under consideration.

For a comparison with Eq. (10.7) it is more advantageous to formulate the scaling in Fourier transforms with respect to the coordinates. Then, we have the scaled microscopical charge density in a simplified notation

$$\hat{\rho}_{int}^{micro}(\lambda^2 t)_\lambda = \hat{\mathcal{H}}_\lambda \left[e^{-\hat{A}_\lambda \lambda^2 t}\left(\hat{\mathcal{P}}(0)_\lambda - \hat{\mathcal{P}}(\infty)_\lambda\right) + \hat{\mathcal{P}}(\infty)_\lambda\right]\,,$$

where the lower index λ means taking the scaled arguments $\frac{1}{\lambda}\vec{k}, \lambda\ell, \lambda\vec{x}_0$. If one compares this with Eq.(10.7), one may remark that though there are formal similarities, in the microscopical solution the spectrum of $\hat{A}_\lambda$ is too rich and therefore it does not give rise to a single exponential dependence.

What happens when we let λ tend to infinity? Generally speaking, the spectrum of $\hat{A}_\lambda$ lies in the right half-plane with the following properties:

i) There is one eigenvalue that goes to zero on the real axis as λ^{-2}.

ii) There might be a non-degenerate eigenvalue that is vanishing for any λ.

iii) The remaining part of the spectrum is bounded from below.

Therefore, for $\lambda \to \infty$ from $e^{-\hat{A}_\lambda \lambda^2 t}$ only the two exponentials corresponding to the two vanishing eigenvalues will survive. (One of them, corresponding to (i) gives rise to a true exponential, while the one corresponding to (ii) is just a constant.)

While this can be defined as the basic mechanism, it also shows that the surviving projections and all other terms in the $\lambda \to \infty$ limit give rise exactly to Eq. (10.7) with definite expressions for D, ϵ, ℓ in terms of the microscopical characteristics.

10.3 Scaling in a neutral hopping model

Before analyzing models with charged particles, we want to discuss here a simple neutral particle model, with the purpose to obtain diffusive behavior through scaling.

Let us consider a system of neutral fermions on a lattice of sites $\vec{x}_i$. The average occupation number of the site i at the instant t is $\bar{n}_i(t)$. The energy of a fermion on the site i is ϵ_i . The total average energy is

$$\bar{E}_n = \sum_i \epsilon_i \bar{n}_i \, .$$

The state of the system is given by the ensemble of average occupation numbers, determined by the hopping rate equation

$$\frac{\partial}{\partial t}\bar{n}_i(t) = -\sum_{i,j} \left[w_{ij}\bar{n}_i(t)(1 - \bar{n}_j(t)) - w_{ji}\bar{n}_j(t)(1 - \bar{n}_i(t)) \right] , \qquad (10.9)$$

where the site-to-site transition rates w_{ij} obey the "detailed balance" relation

$$w_{ij} = w_{ji} e^{\beta(\epsilon_i - \epsilon_j)} \, ,$$

and are supposed to ensure the connectivity of the lattice.

The only microscopical local observable we may define is the average particle density

$$n(\vec{x}, t) = \sum_i \bar{n}_i(t) \delta(\vec{x} - \vec{x}_i) \, . \qquad (10.10)$$

There is no physically meaningful definition for the microscopical current, although many choices are possible.

The general properties of the rate equation were already discussed earlier in Chapter 8. We are again interested only in weak deviations $\delta\bar{n}_i(t)$ of the average occupation numbers from their equilibrium values f_i. The evolution of this deviation is governed by a linear equation. According to the detailed discussion of Chapter 8, the time dependence of the solution is given by

$$\delta\bar{n}(t) = e^{-\Gamma t}\delta\bar{n}(0) \, ,$$

where the matrix Γ_{ij} is built up of the transition rates and the Fermi-function

$$\Gamma_{ij} \equiv \left(\delta_{ij} \sum_l \mathcal{W}_{il} - \mathcal{W}_{ij} \right) \frac{1}{f_i(1 - f_i)}. \tag{10.11}$$

If the initial condition is orthogonal to the only null-eigenvector of Γ, i.e. $\sum_i \delta\bar{n}(0) = 0$, then at $t \to \infty$ the deviations from equilibrium disappear.

On a periodical lattice (not necessarily Bravais-type) one can explicitly implement the scaling procedure we put forward. The positions of the sites in the lattice are given by

$$\vec{x}_{\vec{R},s} = \vec{R} + \vec{\xi}_s; \ (s = 1, \dots S),$$

where $\vec{R}$ runs through the Bravais lattice, while $\vec{\xi}_s$ gives the position of the site within the elementary cell. Translational symmetry implies

$$\epsilon_{\vec{R},s} \equiv \epsilon_s; \ w_{\vec{R},s;\vec{R}',s'} \equiv w_{ss'}(\vec{R} - \vec{R}').$$

It is then natural to use discrete Fourier transforms for any function $f(\vec{R})$ on the lattice defined as

$$\tilde{f}(\vec{k}) = \sum_{\vec{R}} e^{i\vec{k}\vec{R}} f(\vec{R}); \quad f(\vec{R}) = \frac{v}{(2\pi)^3} \int_{BZ} d\vec{k} e^{-i\vec{k}\vec{R}} \tilde{f}(\vec{k}),$$

where v is the volume of the elementary cell and the integration in $\vec{k}$-space is carried out over the Brillouin zone.

Then, for the Fourier transform of $\delta n_s(\vec{R})$ we have

$$\delta\tilde{n}_s(\vec{k}, t) = \sum_{s'=1}^{S} \left(e^{-\tilde{\Gamma}(\vec{k})t} \right)_{ss'} \delta\tilde{n}_s(\vec{k}, 0). \tag{10.12}$$

For each $\vec{k}$ we have a finite-size matrix problem in the cell. The matrix $\tilde{\Gamma}(\vec{k})$ has the following properties:

(i) It is hermitian $\tilde{\Gamma}(\vec{k}) = \tilde{\Gamma}(\vec{k})^+$ in the weighted scalar product

$$(x, y) \equiv \sum_{s=1}^{S} p_s^{-1} x_s^* y_s,$$

with

$$p_s \equiv \frac{f_s(1 - f_s)}{\sum_{s'}^{S} f_{s'}(1 - f_{s'})}.$$

(ii) $\tilde{\Gamma}(\vec{k}) = \tilde{\Gamma}(-\vec{k})^*$ results from the reality of Γ.

(iii) $\tilde{\Gamma}(z\vec{k})$ is analytic in z around $z = 0$, due to the assumed finite range of $w_{ss'}(\vec{R})$.

(iv) $\tilde{\Gamma}(\vec{k}) > 0$ for $\vec{k} \neq 0$ and $\tilde{\Gamma}(0)$ has a single null-eigenvector (by the connectivity argument we discussed in Chapter 8).

Then, it follows that all the eigenvalues $\gamma_\alpha(\vec{k})$ ($\alpha = 1, \ldots, S$) are positive. As $\vec{k}$ goes to zero, all the eigenvalues remain strictly positive with the exception of the lowest one $\gamma_1(\vec{k})$ that has to vanish quadratically, due to (ii) and (iv):

$$\gamma_\alpha(0) > 0; \; \alpha \neq 1,$$

$$\gamma_1(\vec{k}) \sim_{\vec{k} \to 0} \vec{k} \mathcal{D} \vec{k}.$$

The matrix element of the eigenprojector corresponding to the null-eigenvalue at $\vec{k} = 0$ is

$$P^{(1)}(0)_{ss'} = p_s,$$

i.e. not depending on the second index.

Now, we are ready to discuss the scaling limit. We can write Eq.(10.12) in its spectral decomposition form. With scaled parameters it looks:

$$\delta\tilde{n}_s\left(\frac{1}{\lambda}\vec{k}, \lambda^2 t\right) = \sum_{\alpha=1}^{S} e^{-\gamma_\alpha\left(\frac{1}{\lambda}\vec{k}\right)\lambda^2 t} P^{(\alpha)}\left(\frac{1}{\lambda}\vec{k}\right)_{ss'} \delta\tilde{n}_{s'}\left(\frac{1}{\lambda}\vec{k}, 0\right).$$

When λ tends to infinity, only the term $\alpha = 1$ survives

$$\lim_{\lambda \to \infty} \delta\tilde{n}_s\left(\frac{1}{\lambda}\vec{k}, \lambda^2 t\right) = e^{-\vec{k}\mathcal{D}\vec{k}t} p_s \sum_{s'=1}^{S} \delta\tilde{n}_{s'}(0, 0).$$

In order to simulate the fundamental solution of the diffusion equation we must consider an initial condition that can be normalized to unity

$$\sum_{\vec{R}} \sum_{s=1}^{S} \delta n_s(\vec{R},0) = 1$$

or

$$\sum_{s=1}^{S} \delta \tilde{n}_s(0,0) = 1 .$$

This corresponds to putting a supplementary particle at a finite microscopical distance from the origin. Then

$$\lim_{\lambda \to \infty} \delta \tilde{n}_s(\frac{1}{\lambda}\vec{k}, \lambda^2 t) = e^{-\vec{k}\mathcal{D}\vec{k}t} p_s . \tag{10.13}$$

On the other hand, the Fourier transform of the density, defined according to Eq.(10.10) (more exactly, its deviation from the equilibrium value) is given by

$$\delta \tilde{n}(\vec{k},t) = \sum_{s'=1}^{S} \delta \tilde{n}_{s'}(\vec{k},t) .$$

Therefore,

$$\lim_{\lambda \to \infty} \delta \tilde{n}(\frac{1}{\lambda}\vec{k}, \lambda^2 t) = e^{-\vec{k}\mathcal{D}\vec{k}t} ,$$

or in the coordinate space

$$\lim_{\lambda \to \infty} \lambda^3 \delta n(\lambda\vec{x}, \lambda^2 t) = \frac{1}{(4\pi t)^{\frac{3}{2}} [\det \mathcal{D}]^{\frac{1}{2}}} e^{-\frac{\vec{x}\mathcal{D}\vec{x}}{4t}} ,$$

which is exactly the fundamental solution of the diffusion equation with the diffusivity tensor $\mathcal{D}_{\mu\nu}$.

In order, to obtain the solution corresponding to an arbitrary initial condition one has to scale also its geometrical characteristics (to make it macroscopical).

One can see that Eq.(10.13) expresses the Chapman-Enskog limiting situation, where the state depends on time only through the density of particles.

As a side remark, it is useful to retain that the small $\vec{k}$ development of $\Gamma(\vec{k})$ (Moyal expansion) itself would give a wrong result, except on a

Bravais lattice. *The equation itself does not approach the diffusion equation, however, its solution approaches the solution of the diffusion equation.* Physicists tend to ignore that the two statements are not equivalent.

We could discuss here this simple neutral example to full understanding (although we have omitted some steps necessary for full rigor). Unfortunately, the mathematics of charged hopping is much too complicated to include it here. Therefore, in the next section we give only the basic ideas and main results, without any detailed proof.

10.4   Scaling in a charged hopping model

The rate equation for charged hopping fermions in the self-consistent potential approximation looks similar to Eq. (10.9)

$$\frac{\partial}{\partial t}\bar{n}_i(t) = -\sum_{i,j}\left[W_{ij}^t\bar{n}_i(t)(1-\bar{n}_j(t)) - W_{ji}^t\bar{n}_j(t)(1-\bar{n}_i(t))\right], \quad (10.14)$$

however, with time-dependent transition rates W_{ij}^t obeying an instantaneous "detailed balance" relation

$$W_{ij}^t = W_{ji}^t e^{\beta(E_i(t)-E_j(t))}, \quad (10.15)$$

where $E_i(t)$ is the self-consistent energy of the electron on site i at the instant t (determined by the average Coulomb configuration of the system at this instant). It can be decomposed into an unperturbed equilibrium and a non-equilibrium parts:

$$E_i(t) = \varepsilon_i + e\mathcal{V}_i(t), \quad (10.16)$$

where

$$\varepsilon_i = \epsilon_i + e^2 \sum_{j(\neq i)} \frac{f(\varepsilon_j) - q_j}{|\vec{x}_i - \vec{x}_j|}, \quad (10.17)$$

where q_j are the compensating background charges on the sites j and the s.c. potential is determined by the deviation of the charge distribution from equilibrium and the (static) external on-site potential $\mathcal{V}_i^{ext}$

$$\mathcal{V}_i(t) = e \sum_{j(\neq i)} \frac{\bar{n}_j - f(\varepsilon_j)}{|\vec{x}_i - \vec{x}_j|} + \mathcal{V}_i^{ext} \, . \tag{10.18}$$

After a twofold linearization (in $\delta\bar{n}_i(t) \equiv \bar{n}_i(t) - f(\varepsilon_j)$ and $\mathcal{V}_i(t)$) one gets the following equation

$$\frac{d}{dt}\delta\bar{n}_i(t) = - \sum_j \Gamma_{ij} \left[\delta\bar{n}_j(t) + \beta f_j(1 - f_j)e\mathcal{V}_j(t)\right] \, ,$$

where $f_j \equiv f(\varepsilon_j)$ and Γ_{ij} are given by Eq.(10.11) with the equilibrium transition rates w_{ij}, however, with the s.c. equilibrium site energies ε_i. A remarkable property of this result is that it was obtained without knowing the explicit energy-dependence of the transition rates , but merely using the "detailed balance" Eq.(10.15).

Introducing here the expression of the s.c. on-site potential Eq. 10.18) one gets

$$\frac{d}{dt}\delta\bar{n}_i(t) = - \sum_j \left[A_{ij}\delta\bar{n}_j(t) + B_j\mathcal{V}_j^{ext}\right] \, , \tag{10.19}$$

with

$$A_{ij} = \sum_l \Gamma_{il} \left(\delta_{lj} + \beta f_l(1 - f_l)e^2 \frac{1 - \delta_{lj}}{|\vec{x}_l - \vec{x}_j|} \right) \, ,$$

and

$$B_j = -\beta e \Gamma_{ij} f_j(1 - f_j) \, ,$$

also, we got the stipulated general structure. Comparing with the neutral hopping model, we got a modified evolution matrix and a non-homogenous term.

An artefact of the self-consistent potential approximation is that the evolution operator is not always positive. Nevertheless, we shall see that under reasonable conditions stability is granted.

Let us now consider again hopping on a periodical lattice. However, now we shall consider a more complicated connectivity structure than the one considered in the treatment of the neutral model, in order to provide the model with both "free" and "bound" electrons.

The sites within a given cell have a label corresponding to "free" or "bound". The "free" sites form a connected lattice (through the transition rates), while the "bound" sites are connected only between themselves and only within the same cell. Only the Coulomb interaction connects all the sites. Therefore charge may flow on the "free" sites, while on the "bound" sites only a local polarization (within the cell) may occur.

We shall define the corresponding electromagnetic lengths by the respective densities

$$\frac{1}{\ell^2} = 4\pi e^2 \frac{\partial n^0_{free}}{\partial \mu} \; ; \qquad \frac{1}{L^2} = 4\pi e^2 \frac{\partial n^0_{bound}}{\partial \mu} \; .$$

Equation (10.19) on the periodical lattice after a discrete Fourier transform, looks as

$$\frac{d}{dt}|\delta\tilde{n}(\vec{k},t)\rangle = -\tilde{A}(\vec{k})|\delta\tilde{n}(\vec{k},t)\rangle - \frac{v}{e}\tilde{\Gamma}(\vec{k})\nu|\tilde{\mathcal{V}}^{ext}(\vec{k})\rangle \, , \qquad (10.20)$$

where an operator and ket notation (in the cell-space) was used for convenience and

$$\nu_s \equiv \frac{\beta e^2}{v} f_s(1 - f_s)\delta_{ss'} \; .$$

One may also write

$$\tilde{A}(\vec{k}) = \tilde{\Gamma}(\vec{k})\left(1 + \nu\tilde{C}(\vec{k})\right) \, ,$$

where $\tilde{C}(\vec{k})$ is the discrete Fourier transform of the Coulomb matrix. It is completely determined by

$$C_{ss'}(\vec{r}) \equiv \frac{v\left(1 - \delta_{ss'}\delta_{\vec{r}0}\right)}{|\vec{r} + \vec{\xi}_s - \vec{\xi}_{s'}|} = \frac{v}{(2\pi)^3}\int_{BZ} d\vec{k}\, e^{-i\vec{k}\vec{r}}\tilde{C}_{ss'}(\vec{k}) \, ,$$

although the inverse transform does not exist in the ordinary sense. It can be shown that $\tilde{C}_{ss'}(\vec{k})$ has the following structure:

$$\tilde{C}_{ss'}(\vec{k}) = \frac{1}{k^2}e^{-i\vec{k}(\vec{\xi}_s - \vec{\xi}_{s'})} + \tilde{R}_{ss'}(\vec{k}) \, ,$$

where $\tilde{R}_{ss'}(\epsilon\vec{k})$ is analytical in ϵ around $\epsilon = 0$.

Using an old result of Onsager

$$C \geq -\frac{v}{2d},$$

with d being the smallest distance in the lattice, and choosing

$$\ell, L > \sqrt{\frac{v}{2d}},$$

we can ensure the positivity of the matrix

$$\nu^{-\frac{1}{2}}\tilde{X}\nu^{\frac{1}{2}} \equiv 1 + \nu^{\frac{1}{2}}\tilde{C}\nu^{\frac{1}{2}}.$$

Then it it is useful to write the solution of Eq.(10.20) as

$$|\delta\tilde{n}(\vec{k},t)\rangle = e^{-\tilde{A}(\vec{k})t}\left[|\delta\tilde{n}(\vec{k},0)\rangle + \frac{v}{e}\tilde{X}(\vec{k})^{-1}\nu|\tilde{\mathcal{V}}^{ext}(\vec{k})\rangle\right] - \frac{v}{e}\tilde{X}(\vec{k})^{-1}\nu|\tilde{\mathcal{V}}^{ext}(\vec{k})\rangle.$$

Now, the following properties of the evolution operator $\tilde{A}(\vec{k})$ can be shown:

a) $\tilde{A}(\vec{k})$ is diagonalizable, with real, positive eigenvalues bounded as functions of $\vec{k} \in BZ$. It has a null-eigenvector related to the existence of "bound" electrons. (These properties are not trivial, since $\tilde{A}(\vec{k})$ is neither hermitian nor bounded.)

b) The behavior of the spectrum and eigenprojectors of $\tilde{A}_\lambda \equiv \tilde{A}(\frac{1}{\lambda}\vec{k}, \lambda\ell)$ is an analytic perturbation problem. The eigenvalues are analytic in λ^{-1} around the origin, while the eigenprojectors may have poles. There is just one eigenvalue that goes to zero from above as λ^{-2}.

The scaling of ℓ is an essential ingredient. With the exception of the zero eigenvalue due to the "bound electrons" all the other eigenvalues of $\tilde{A}(\vec{k})$ remain positive, even for $\vec{k} = 0$.

Now, if we consider neutral islands of "bound electrons"

$$\sum_{s\in bound} \delta\bar{n}_s(\vec{r},t) = 0,$$

an initial state with a supplementary "free" electron in the cell $\vec{r} = 0$

$$\sum_{s \in free} \delta \bar{n}_s(\vec{r}, 0) = \delta_{\vec{r}, 0}$$

and a point external charge q in a point $\vec{x}_0$ (not belonging to the lattice), i.e. an external potential

$$V_s^{ext}(\vec{r}) = \frac{q}{|\vec{r} + \vec{\xi}_s - \vec{x}_0|},$$

one may rigorously show that

$$\lim_{\lambda \to \infty} \hat{\rho}_T^{micro}(\frac{1}{\lambda}\vec{k}, \lambda^2 t; \lambda \ell, \lambda \vec{x}_0) = \hat{\rho}_T(\vec{k}, t; \ell, \vec{x}_0),$$

with the diffusivity tensor $D_{\mu\nu}$ determined completely by the entities pertinent to the "free" electrons corresponding to the previously discussed neutral model, and the dielectric permittivity tensor $\epsilon_{\mu\nu}$ determined completely by the entities pertinent to the "bound" electrons.

While the diffusivity is given again by the curvature at $\vec{k} = 0$ of the lowest eigenvalue of the Γ matrix (restricted to the "free" electrons), the dielectric permittivity is given by the scalar product (matrix element)

$$\epsilon_{\mu\nu} = \delta_{\mu\nu} + \frac{1}{L^2}\langle \xi_\mu | \sqrt{q}\Pi \left[1 + \frac{1}{L^2}\Pi\sqrt{q}\tilde{R}(0)\sqrt{q}\Pi \right]^{-1} \Pi\sqrt{q}|\xi_\nu\rangle,$$

where the operator Π is defined by

$$\Pi \equiv 1 - |\sqrt{q}\rangle\langle\sqrt{q}|,$$

and

$$q_s \equiv \frac{f_s(1 - f_s)}{\sum_{s' \in bound} f_{s'}(1 - f_{s'})},$$

with s belonging to the "bound" electron sites.

The "free" electrons provide the possibility for a macroscopic current flow, while the "bound" electrons give the local polarization properties.

It can be seen from the detailed analysis that while the polarization charge does not set on instantaneously as in the macroscopic description, the respective time scale is microscopical (vanishing after scaling).

The resulting dielectric permittivity contains automatically also the local field effects, and it can be shown that for a cubic lattice the Clausius-Mosotti formula

$$\epsilon - 1 = \frac{\kappa}{1 - \frac{1}{3}\kappa}$$

holds, with κ being the polarizability of the cell. This relationship, however, has to be understood in the sense of equality of asymptotic series in the inverse of the lattice constant. This last result may be extended to the general quantum mechanical theory of dielectric permittivity of crystals [Bányai and Gartner (1984)].

10.5 Scaling in a charged Boltzmann model

In this Section we will consider the scaling procedure in the kinetic model of charged particles described by the Boltzmann equation for the distribution function of coordinates and velocities $p(\vec{v}, \vec{x}, t)$:

$$\left(\frac{\partial}{\partial t} + \vec{v}\frac{\partial}{\partial \vec{x}} + \frac{e}{m}\vec{E}(\vec{x}, t)\frac{\partial}{\partial \vec{v}}\right) p(\vec{v}, \vec{x}, t)$$

$$= -\int d\vec{v}' \left[w(\vec{v}, \vec{v}')p(\vec{v}, \vec{x}, t) - w(\vec{v}', \vec{v})p(\vec{v}', \vec{x}, t)\right] . \quad (10.21)$$

The instant thermalization model is defined by the transition rates

$$w(\vec{v}, \vec{v}') = e^{\beta \frac{m}{2}(v^2 - v'^2)}w(\vec{v}', \vec{v}) = \frac{1}{\tau}\left(\frac{m\beta}{2\pi}\right)^{\frac{3}{2}} e^{-\beta \frac{mv'^2}{2}} .$$

In the special case of full homogeneity (homogenous electric field and homogeneous initial conditions) we can use the normalization condition $V\int d\vec{v}p(\vec{v}, t) = 1$ and thereafter the Boltzmann equation looks as

$$\left(\frac{\partial}{\partial t} + \frac{e}{m}\vec{E}(t)\frac{\partial}{\partial \vec{v}}\right) p(\vec{v}, \vec{x}, t) = -\frac{1}{\tau}\left(p(\vec{v}, t) - p_0(v)\right) .$$

Here $p_0(v)$ is the normalized Maxwell-Boltzmann distribution

$$p_0(v) \equiv \frac{1}{V} \left(\frac{m\beta}{2\pi} \right)^{\frac{3}{2}} e^{-\beta \frac{mv^2}{2}} .$$

If the electric field is harmonically oscillating $\vec{E}(t) = \Re \left\{ \vec{E} e^{\imath \omega t} \right\}$, then the stationary oscillating solution, up to terms of first order in the field is

$$p(\vec{v}) = p_0(v) - \Re \left\{ \frac{1}{\imath \omega + \tau^{-1}} \vec{E} e^{\imath \omega t} \right\} \frac{e}{m} \frac{\partial}{\partial \vec{v}} p_0(v) ,$$

and the current density is

$$\vec{j}(t) = Ne \int d\vec{v} p(\vec{v}, t) \vec{v} = -\Re \left\{ \frac{1}{\imath \omega + \tau^{-1}} \vec{E} e^{\imath \omega t} \right\} \frac{Ne^2}{m} \int d\vec{v} \vec{v} \frac{\partial}{\partial \vec{v}} p_0(v)$$

$$= \frac{n_0 e^2}{m} \Re \left\{ \frac{1}{\imath \omega + \tau^{-1}} \vec{E} e^{\imath \omega t} \right\} .$$

Therefore, one gets for the complex conductivity the Drude formula

$$\sigma(\omega) = \frac{n_0 e^2}{m} \frac{1}{\imath \omega + \tau^{-1}} .$$

From its low-frequency real and imaginary parts one identifies

$$\sigma = \frac{n_0 e^2}{m} \tau, \ \epsilon - 1 = \frac{4\pi n_0 e^2}{m} \tau^2 . \tag{10.22}$$

Using the Einstein relation Eq.(10.1 the diffusion constant D results as

$$D = \frac{k_B T}{m} \tau . \tag{10.23}$$

An apparent paradox of these identifications is the non-trivial dielectric permittivity in a model containing only "free" charges. We comment on it later, after the full analysis of the model.

Obviously, in the homogeneous case no internal field appears due the flowing particles.

The more interesting case for our purpose is the inhomogeneous transient flow in the presence of a static external charge. We ignore the internal magnetic field. The electric field is then determined self-consistently from the Poisson equation

$$\nabla \vec{E}(\vec{x}, t) = 4\pi e N \int d\vec{v} \left(p(\vec{v}, \vec{x}, t) - p_0(v) \right) + 4\pi \rho_{ext}(\vec{x}),$$

by the external charge density and the instantaneous deviation of the average internal charge density from its constant equilibrium value. (A compensating uniform background charge of density $n_0 = \frac{N}{V}$ is understood.)

In this model all the "electrons" are "free charges" and therefore

$$\frac{1}{\ell^2} = \frac{4\pi e^2 n_0}{k_B T}.$$

For small deviations from equilibrium and weak electric fields one can write down the linearized equations for the deviation of the distribution from the equilibrium one ($\eta \equiv p - p_0$):

$$\left(\frac{\partial}{\partial t} + \vec{v} \frac{\partial}{\partial \vec{x}} \right) \eta(\vec{v}, \vec{x}, t) + \frac{e}{m} \vec{E}(\vec{x}, t) \frac{\partial p_0}{\partial \vec{v}}$$

$$= -\frac{1}{\tau} \left[\eta(\vec{v}, \vec{x}, t) - V p_0(v) \int d\vec{v}' \eta(\vec{v}', \vec{x}, t) \right],$$

and

$$\nabla \vec{E}(\vec{x}, t) = 4\pi e N \int d\vec{v} \eta(\vec{v}, \vec{x}, t) + 4\pi \rho_{ext}(\vec{x}).$$

After a Fourier transformation in the coordinate, these equations may be combined into a single equation for the Fourier transform $\tilde{\eta}(\vec{v}, \vec{k}, t) \equiv \int d\vec{x} e^{-i\vec{k}\vec{x}} \eta(\vec{v}, \vec{x}, t)$

$$\frac{\partial}{\partial t} \tilde{\eta}(\vec{v}, \vec{k}, t) = - \left(i\vec{k}\vec{v} + \tau^{-1} \right) \tilde{\eta}(\vec{v}, \vec{k}, t)$$

$$+ \left(\tau^{-1} - \frac{i4\pi \vec{k}\vec{v}}{\ell^2 k^2} \right) V p_0(v) \int d\vec{v}' \tilde{\eta}(\vec{v}', \vec{k}, t)$$

$$- \frac{i4\pi \vec{k}\vec{v}}{\ell^2 k^2 e n_0} V p_0(v) \tilde{\rho}_{ext}(\vec{k}), \tag{10.24}$$

or in operator notations

$$\frac{\partial}{\partial t}\tilde{\eta} = -\tilde{A}\tilde{\eta} + \tilde{b},$$ (10.25)

with the matrix elements

$$\tilde{A}(\vec{k})_{\vec{v}\vec{v}'} \equiv \left(\imath\vec{k}\vec{v} + \tau^{-1}\right)\delta(\vec{v}-\vec{v}') - \left(\tau^{-1} - \frac{\imath 4\pi\vec{k}\vec{v}}{\ell^2 k^2}\right)Vp_0(v),$$

and the column array

$$\tilde{b}(\vec{k})_{\vec{v}} \equiv -\frac{\imath 4\pi\vec{k}\vec{v}}{\ell^2 k^2 e n_0}Vp_0(v)qe^{\imath\vec{k}\vec{x}_0}.$$

Here we explicitly introduced a point external charge q in the point $\vec{x}_0$

$$\rho_{ext}(\vec{x}) = q\delta(\vec{x} - \vec{x}_0).$$

We consider also an initial condition that would correspond to the macroscopic fundamental solution

$$\eta(\vec{v}, \vec{x}, 0) = \delta(\vec{x})g(\vec{v}),$$

with an arbitrary normalized function of velocity $g(\vec{v})$ ($\int d\vec{v}g(\vec{v}) = 1$). Since we are already in a linear theory the smallness of η plays no role. In the same time, the fact that an accumulation of particles in the point $\vec{x} = 0$ has to be compensated by a homogeneous depletion in the whole volume, may be ignored in the limit of the large volume.

It is convenient to work out this problem in Laplace transforms with respect to the time variable :

$$\hat{\tilde{\eta}}(s) \equiv \int_0^\infty dt e^{-st}\tilde{\eta}(t).$$

From Eq. (10.25) we get immediately

$$\hat{\tilde{\eta}}(s) = \frac{1}{s + \tilde{A}}\left(\tilde{\eta}(0) + \frac{1}{s}\tilde{b}\right).$$ (10.26)

Now, the evolution operator $\tilde{A}$ can be split into two pieces

$$\tilde{A} = T(\vec{k}) - J(\vec{k}, \ell),$$

with a diagonal (unperturbed) part

$$T(\vec{k})_{\vec{v}\vec{v}'} \equiv \left(\imath \vec{k}\vec{v} + \tau^{-1}\right) \delta(\vec{v} - \vec{v}'),$$

and a non-diagonal (perturbation) part

$$J(\vec{k}, \ell)_{\vec{v}\vec{v}'} \equiv \psi(\vec{k}, \vec{v}, \ell),$$

with

$$\psi(\vec{k}, \vec{v}, \ell) \equiv \left(\tau^{-1} - \frac{\imath 4\pi \beta \vec{k}\vec{v}}{\ell^2 k^2}\right) V p_0(v).$$

As it is easy to see

$$J^2 = \frac{1}{\tau} J,$$

i.e. τJ is an one-dimensional projector.

Let us consider the resolvent operator

$$R(z) \equiv \frac{1}{z + A}$$

that appears in Eq.(10.26). Furthermore we need only to find the explicite form of the resolvent. On this purpose we express it through the resolvent $R_0 = \frac{1}{z+T}$ of the diagonal part of A

$$R(z) \equiv \frac{1}{z + A} = R_0(z)\frac{1}{1 - JR_0(z)},$$

having the formal Taylor expansion

$$R = R_0 \left(1 + JR_0 + JR_0 JR_0 + JR_0 JR_0 JR_0 + \cdots\right).$$

On the other hand, we have

$$JR_0 J = \mathcal{F}(z, \vec{k})J,$$

with

$$\mathcal{F}(z; \vec{k}, \ell) \equiv \int d\vec{v}\, \psi(\vec{k}, \vec{v}, \ell) \frac{1}{z + \frac{1}{\tau} + \imath\beta\vec{k}\vec{v}} \,.$$

Using this relation and the projector property of τJ one may resum the series getting

$$R(z) = R_0(z) + \frac{1}{1 - \mathcal{F}(z)} R_0(z) J R_0(z) \,. \qquad (10.27)$$

Thus, one may conclude that the perturbed spectrum consists of the unperturbed one (a cut defined by $\Re z = -\tau^{-1}$) and poles determined by the zeroes of the function $1 - \mathcal{F}(z)$. Again it may be shown that for small k and large ℓ one of these zeroes goes to the origin on the real axis coming from positive values.

Let us examine directly the scaling limit of the solution in Laplace transforms.

Using the expansion

$$\frac{1}{\frac{1}{\lambda^2}s + \frac{1}{\tau} + \frac{\imath}{\lambda}\vec{k}\vec{v}} = \tau\left[1 - \frac{\imath}{\lambda}\vec{k}\vec{v}\tau - \frac{1}{\lambda^2}\left(s\tau + \left(\beta\vec{k}\vec{v}\right)^2 \tau^2\right) + \dots\right],$$

it is easy to show that

$$\lim_{\lambda\to\infty} \lambda^2\left(1 - F\left(\frac{1}{\lambda^2}s; \frac{1}{\lambda}\vec{k}, \lambda\ell\right)\right) = \left[\left(1 + \frac{1}{k^2\ell^2}\right)Dk^2 + s\right]\tau \,.$$

Then, after a simple algebra one finds

$$\lim_{\lambda\to\infty} \lambda^2\hat{\tilde{\eta}}\left(\frac{1}{\lambda^2}s, \frac{1}{\lambda}\vec{k}, \lambda\ell, \lambda\vec{x}_0; \vec{v}\right) = \frac{\frac{1}{n_0}p_0(v)}{\left(1 + \frac{1}{k^2\ell^2}\right)Dk^2 + s}\left(1 - \frac{Dq}{s\ell^2 e}e^{\imath\vec{k}\vec{x}_0}\right),$$

or for the scaled Laplace and Fourier transformed internal charge density

$$\lim_{\lambda\to\infty} \lambda^2 eN \int d\vec{v}\,\hat{\tilde{\eta}}\left(\frac{1}{\lambda^2}s, \frac{1}{\lambda}\vec{k}, \lambda\ell, \lambda\vec{x}_0; \vec{v}\right)$$
$$= \frac{1}{\left(1 + \frac{1}{k^2\ell^2}\right)Dk^2 + s}\left(e - \frac{Dq}{s\ell^2}e^{\imath\vec{k}\vec{x}_0}\right).$$

It can be easily seen that this is indeed the Laplace transform in time of the macroscopic (internal) charge density Eq. (10.7) for $\epsilon = 1$. Therefore, according to our scaling analysis that covers the general solution of the problem, there is no polarization charge in this model and the permittivity is that of the vacuum. The contradiction with the second of Eqs. (10.22) is only apparent. Actually, that interpretation gave rise to a susceptibility $\frac{ne^2}{m}\tau^2 = \frac{1}{3}\left(\frac{l}{\ell}\right)^2$, where $l = \tau\sqrt{\langle v^2\rangle_0}$ is the mean free path and therefore in the scaling limit as $\ell \to \infty$ it disappears. This is also physically correct, since $\ell \gg l$ in a semiconductor. Only within this precision there is a unique ϵ in all possible experiments, differing by terms that vanish when $\ell \to \infty$.

*

We could see in this chapter, on the example of solvable models, that the macroscopical hydrodynamic behavior emerges in the large time and large space scale limits rigorously from the solution of the underlying kinetic theory. We believe in the generality of this result. On the other hand, we were able to cover only linear phenomena, where the necessary mathematical tools are available.

Chapter 11

Linear response to external perturbations

We have seen in the previous chapters the way one may determine the kinetic coefficients within the frame of some solvable models. In this chapter we want to try a more general approach for the determination of the same coefficients.

Let us have a system of particles described by the Hamiltonian H. We want to test the reaction of this system to a weak external field [Kubo (1957); Greenwood (1958)]. In this case, the total Hamilton operator includes also the time-dependent interaction with this external field

$$\mathcal{H}(t) = H + H'(t).$$

The time evolution is described by the quantum mechanical Liouville equation (von Neumann equation) for the density operator:

$$i\hbar \frac{\partial \rho}{\partial t} = [\mathcal{H}(t), \rho]. \tag{11.1}$$

We choose as initial state at the time t_0 the grand canonical ensemble:

$$\rho(t_0) = \rho_0(H) \equiv \frac{e^{-\beta(H-\mu N)}}{Z}.$$

In the interaction picture:

$$\bar{\rho}(t) = e^{\frac{i H(t-t_0)}{\hbar}} \rho(t) e^{-\frac{i H(t-t_0)}{\hbar}},$$

one gets the Liouville equation

$$i\hbar \frac{\partial \bar{\rho}}{\partial t} = [\bar{H}'(t), \bar{\rho}]; \qquad \bar{H}'(t) = e^{\frac{i H(t-t_0)}{\hbar}} H'(t) e^{-\frac{i H(t-t_0)}{\hbar}}.$$

The equation for the deviation from equilibrium

$$\bar{\rho} = \rho_0 + \delta\bar{\rho},$$

is

$$i\hbar\frac{\partial\delta\bar{\rho}(t)}{\partial t} = \left[\bar{H}'(t), \rho_0\right] + \dots,$$

with

$$\delta\bar{\rho}(t_0) = 0,$$

and the points indicate terms of higher order in the external perturbation. The formal solution is

$$\delta\bar{\rho}(t) = \frac{1}{i\hbar}\int_{t_0}^{t} dt' \left[\bar{H}'(t'), \rho_0\right] + \dots,$$

and

$$\rho(t) = \rho_0 + \frac{1}{i\hbar}\int_{t_0}^{t} dt' e^{\frac{iH(t'-t)}{\hbar}} \left[H'(t'), \rho_0\right] e^{\frac{-iH(t'-t)}{\hbar}} + \dots,$$

where only the lowest order terms are written down explicitly.

The average of an observable A in the linear approximation is then

$$\langle A(t)\rangle - \langle A\rangle_0 = \frac{1}{i\hbar}\int_{t_0}^{t} dt' Tr\left\{ e^{\frac{iH(t'-t)}{\hbar}} \left[H'(t'), \rho_0\right] e^{\frac{-iH(t'-t)}{\hbar}} A \right\}$$

$$= \frac{1}{i\hbar}\int_{t_0}^{t} dt' \langle [A_H(t-t'), H'(t')]\rangle_0. \tag{11.2}$$

We have used here the cyclic permutability under the trace and denoted with $\langle\dots\rangle_0$ the equilibrium average.

Let us consider now the example of the interaction with a longitudinal electromagnetic field (defined by a scalar potential $V^{ext}(\vec{x}, t)$)

$$H'(t) = \int d\vec{x}\rho(\vec{x})V^{ext}(\vec{x}, t),$$

and we are looking for the average charge density. According to the above derived formulae

$$\langle \rho(\vec{x}, t) \rangle = \langle \rho(\vec{x}) \rangle_0 + \frac{1}{i\hbar} \int_{t_0}^{t} dt' \int d\vec{x}' \langle [\rho(\vec{x}, t - t')_H , \rho(\vec{x}')] \rangle_0 V^{ext}(\vec{x}', t') ,$$

or after the extension of the time-integral till $-\infty$

$$\langle \rho(\vec{x}, t) \rangle = \langle \rho(\vec{x}) \rangle_0 + \frac{1}{i\hbar} \int_{-\infty}^{t} dt' \int d\vec{x}' \langle [\rho(\vec{x}, t)_H , \rho(\vec{x}', t')_H] \rangle_0 V^{ext}(\vec{x}', t') .$$

We can note that - as it could be expected - a spatially constant potential has no effect, since the total charge is conserved i.e. $Q \equiv \int d\vec{x}\rho(\vec{x})$ commutes with H, H' and $\rho(\vec{x})$.

Due to the same properties, the total charge will not change, due to the introduction of the interaction

$$\langle Q(t) \rangle = \langle Q \rangle_0 .$$

For the sake of simplicity let us suppose that the system in equilibrium was everywhere neutral in space. This means $\langle \rho(\vec{x}) \rangle_0 = 0$. Then we may write formally

$$\langle \rho(\vec{x}, t) \rangle = \int_{-\infty}^{\infty} dt' \int d\vec{x}' \kappa(\vec{x}, \vec{x}'; t - t') V^{ext}(\vec{x}', t') ,$$

with a kernel given by the charge density correlation

$$\kappa(\vec{x}, \vec{x}'; t) \equiv \frac{\theta(t)}{i\hbar} \langle [\rho(\vec{x}, t)_H , \rho(\vec{x}', 0)_H] \rangle_0 .$$

Thus, the charge density has a nonlocal space- and time- dependence on the potential. Nevertheless, since real systems have a finite correlation length, i.e. $\kappa(\vec{x}, \vec{x}'; t)$ vanishes rapidly with the distance $|\vec{x} - \vec{x}'|$, by slowly varying potentials the relationship may be treated in a very good approximation as being local. (However, it is not granted that a given model ensure such short range correlations.) One can notice that the relationship in the time variable is retarded (only the past matters).

An interesting case is that of spatially homogeneous systems, where the kernel depends only on $\vec{x} - \vec{x}'$ i.e. $\kappa(\vec{x}, \vec{x}'; t) \equiv \kappa(\vec{x} - \vec{x}'; t)$. After a Fourier transformation in space and time we then get

$$\langle \tilde{\rho}(\vec{k}, \omega) \rangle = \tilde{\kappa}(\vec{k}, \omega) \tilde{V}^{ext}(\vec{k}, \omega) \,,$$

with

$$\tilde{\kappa}(\vec{k}, \omega) = \int_0^\infty dt \int d\vec{x} e^{-i\vec{k}\vec{x}} e^{i\omega t} \kappa(\vec{x}, t) \,.$$

In a Coulomb interacting system of charged particles the charge density induced by the external field produces a potential by itself. On the other hand, the phenomenological relation we are looking for, is the response to the total potential V. We may get this potential from the Poisson equation, which in Fourier transforms looks as

$$k^2 \tilde{V}(\vec{k}, \omega) = 4\pi \langle \tilde{\rho}(\vec{k}, \omega) \rangle + 4\pi \rho^{ext}(\vec{k}, \omega) \,.$$

Here the external charge is related to the external potential also by a Poisson equation

$$k^2 \tilde{V}^{ext}(\vec{k}, \omega) = 4\pi \rho^{ext}(\vec{k}, \omega) \,.$$

Therefore,

$$k^2 \tilde{V}(\vec{k}, \omega) = 4\pi \tilde{\kappa}(\vec{k}, \omega) \tilde{V}^{ext}(\vec{k}, \omega) + 4\pi \rho^{ext}(\vec{k}, \omega) \,,$$

or

$$\epsilon(\vec{k}, \omega) k^2 \tilde{V}(\vec{k}, \omega) = 4\pi \rho^{ext}(\vec{k}, \omega) \,,$$

with the (longitudinal) dielectric function

$$\epsilon(\vec{k}, \omega) \equiv \frac{1}{1 + \frac{4\pi}{k^2} \tilde{\kappa}(\vec{k}, \omega)} \,.$$

However. if we consider the Coulomb-system in the Hartree approximation, one has to reconsider the definitions. In that case the perturbation is the self-consistent potential itself (which coincides with the total one), not the external potential. Therefore one has

$$\epsilon(\vec{k}, \omega)^{sc} \equiv 1 - \frac{4\pi}{k^2} \tilde{\kappa}^{sc}(\vec{k}, \omega) \, .$$

This distinction is very important and as we shall see later, bears important consequences.

Chapter 12

The Nyquist theorem

The classical Nyquist theorem was deduced [Nyquist (1928)] within the classical thermodynamics of electric circuitry. It is a statement relating the thermal fluctuation of the potential drop U in a conductor to its resistance R and temperature T:

$$\langle \delta U^2 \rangle_0 = R k_B T .$$

However, it can be brought to another form, which is more advantageous from the theoretical point of view. Since the potential drop along the conductor is given by the field strength $\mathcal{E}$ multiplied by the length L of the wire $U = \mathcal{E} L$ and the resistance is related to the d.c. conductance σ_{dc}, the length L and cross section S of the wire by $R = \frac{L}{\sigma_{dc} S}$, one gets

$$\langle \delta \mathcal{E}^2 \rangle_0 = \frac{k_B T}{\Omega \sigma_{dc}} , \tag{12.1}$$

where $\Omega = LS$ is the volume of the wire. This later form is more suitable for a microscopical treatment.

The spectral density of the equilibrium noise of the (hermitian) observable A (whose equilibrium average is vanishing)[1] is defined as:

$$\delta A_\omega^2 \equiv \frac{1}{2} \int_0^\infty dt \cos \omega t \langle \{ A A(t) + A(t) A \} \rangle_0 , \tag{12.2}$$

---

[1] The symmetric product in the definition of the correlation function Eq.(12.2) yields the usual form of the dissipation-fluctuation theorem (DFT) and the Nyquist formula. Correlation functions of the asymmetric products $A A(t)$ and $A(t) A$ yield other forms of the DFT. They differ both from each other and from the usual form with respect to the contribution of vacuum energy fluctuations (see [Brenig (1989)]). For a discussion of the relevance of the vacuum fluctuations in this context the reader is referred to [MacDonald (1962)].

with the notations

$$A(t) = e^{\frac{i}{\hbar}Ht} A e^{-\frac{i}{\hbar}Ht}; \quad \langle ... \rangle_0 \equiv Tr\left(\frac{1}{Z} e^{-\beta(H-\mu N)} ... \right).$$

Here we have assumed that ergodicity holds, in the sense of the vanishing of correlations at infinite time.

The above expression can be written as

$$\delta A_\omega^2 = \frac{1}{2}\Re\left\{\int_0^\infty dt e^{-i\omega t} \langle \{A A(t) + A(t)A\} \rangle_0 \right\}$$

$$= \frac{1}{2Z}\sum_{n,n'}\Re\left\{\int_0^\infty dt e^{-i(\omega-i0)t} |A_{nn'}|^2 e^{-\beta(E_n-\mu N_n)} \right.$$

$$\left. \times \left(e^{-\frac{i}{\hbar}(E_n-E_{n'})t} + e^{\frac{i}{\hbar}(E_n-E_{n'})t}\right)\right\}$$

$$= \frac{\pi}{2Z}\sum_{n,n'}\Re\left\{|A_{nn'}|^2 e^{-\beta(E_n-\mu N_n)} \right.$$

$$\left. \times \left(\delta\left(\omega+\frac{1}{\hbar}(E_n-E_{n'})\right) + (\omega \to -\omega)\right)\right\}$$

$$= \frac{\pi}{2Z}\frac{1+e^{-\beta\hbar\omega}}{1-e^{-\beta\hbar\omega}}\sum_{n,n'}\Re\left\{|A_{nn'}|^2 e^{-\beta(E_n-\mu N_n)} \right.$$

$$\left. \times \left(\delta\left(\omega+\frac{1}{\hbar}(E_n-E_{n'})\right) - (\omega \to -\omega)\right)\right\}$$

$$= \frac{1}{2}\coth\left(\frac{\beta\hbar\omega}{2}\right)\Re\left\{\int_0^\infty dt e^{-i\omega t} \langle [A, A(t)] \rangle_0 \right\}.$$

The microscopical field (operator) is

$$\vec{\mathcal{E}}(\vec{x}, t) = -\nabla V(\vec{x}, t),$$

but

$$V(\vec{x}, t) = V_{ext}(\vec{x}, t) + \int d\vec{x}' \frac{\rho(\vec{x}')}{|\vec{x} - \vec{x}'|},$$

with the operator of the particle density $\rho(\vec{x}, t)$. Since the external field is considered as a commuting c-number, the fluctuation of the field is completely determined by the internal field

$$-\nabla \int d\vec{x}' \, \frac{\rho(\vec{x}')}{|\vec{x} - \vec{x}'|} \, .$$

After a Fourier transform in space

$$V(\vec{k}, t) \equiv \int d\vec{x} e^{-\imath \vec{k}\vec{x}} V(\vec{x}, t) \, ,$$

one has

$$\vec{\mathcal{E}}(\vec{k}, t) = \imath \vec{k} V(\vec{k}, t) = \imath \vec{k} \left(\tilde{V}_{ext}(\vec{k}, t) + \frac{4\pi}{k^2} \tilde{\rho}(\vec{k}, t) \right) \, .$$

From the hermiticity of the field operator follows that $\vec{\mathcal{E}}(\vec{k}, t)^+ = \vec{\mathcal{E}}(-\vec{k}, t)$. The fluctuating part of the field is $\delta\vec{\mathcal{E}}(\vec{k}, t) = \imath \vec{k} \frac{4\pi}{k^2} \rho(\vec{k}, t)$.

Let us now define

$$\langle\langle \delta\vec{\mathcal{E}}(\vec{k}) \delta\vec{\mathcal{E}}(-\vec{k}) \rangle\rangle_\omega \equiv \frac{1}{2} \coth\left(\frac{\beta\hbar\omega}{2} \right) \Re \left\{ \int_0^\infty dt e^{-\imath\omega t} \langle [\delta\vec{\mathcal{E}}(\vec{k}), \delta\vec{\mathcal{E}}(-\vec{k}, t)] \rangle_0 \right\} \, ,$$

(scalar product of vectors is understood) or

$$\langle\langle \delta\vec{\mathcal{E}}(\vec{k}) \delta\vec{\mathcal{E}}(-\vec{k}) \rangle\rangle_\omega \equiv \frac{1}{2} \coth\left(\frac{\beta\hbar\omega}{2} \right) \frac{(4\pi)^2}{k^2} \Re \left\{ \int_0^\infty dt e^{-\imath\omega t} \langle [\tilde{\rho}(\vec{k}), \tilde{\rho}(-\vec{k}, t)] \rangle_0 \right\} \, .$$

On the other hand

$$\langle [\tilde{\rho}(\vec{k}), \tilde{\rho}(-\vec{k}, t)] \rangle_0 = \int d\vec{x} e^{-\imath \vec{k}\vec{x}} \int d\vec{x}' e^{\imath \vec{k}\vec{x}'} \langle [\rho(\vec{x}), \delta\rho(\vec{x}', t)] \rangle_0$$

$$= \int d\vec{x} e^{-\imath \vec{k}\vec{x}} \int d\vec{x}' e^{\imath \vec{k}\vec{x}'} \langle [\rho(\vec{x} - \vec{x}'), \delta\rho(0, t)] \rangle_0$$

$$= \int d\vec{x}' \int d\vec{x} e^{-\imath \vec{k}\vec{x}} \langle [\rho(\vec{x}), \delta\rho(0, t)] \rangle_0$$

$$= \Omega \int d\vec{x} e^{-\imath \vec{k}\vec{x}} \langle [\rho(\vec{x}), \delta\rho(0, t)] \rangle_0 \, ,$$

holds. Therefore,

$$\langle\langle\delta\vec{\mathcal{E}}(\vec{k})\delta\vec{\mathcal{E}}(-\vec{k})\rangle\rangle_\omega \equiv \coth\left(\frac{\beta\hbar\omega}{2}\right)\frac{(4\pi)^2}{2k^2}\Omega$$

$$\times\Re\left\{\int_0^\infty dt e^{-\imath\omega t}\int d\vec{x}e^{-\imath\vec{k}\vec{x}}\langle[\rho(\vec{x}),\delta\rho(0,t)]\rangle_0\right\},$$

or

$$\langle\langle\delta\vec{\mathcal{E}}(\vec{k})\delta\vec{\mathcal{E}}(-\vec{k})\rangle\rangle_\omega \equiv \coth\left(\frac{\beta\hbar\omega}{2}\right)\frac{(4\pi)^2}{2k^2}\Omega\Re\left\{\imath\hbar\tilde{\kappa}(\vec{k},\omega)\right\},$$

results.

Using the relationship

$$\frac{4\pi}{k^2}\tilde{\kappa}(\vec{k},\omega) = \frac{1}{\epsilon(\vec{k},\omega)} - 1,$$

we get

$$\langle\langle\delta\vec{\mathcal{E}}(\vec{k})\delta\vec{\mathcal{E}}(-\vec{k})\rangle\rangle_\omega \equiv -\frac{1}{2}\coth\left(\frac{\beta\hbar\omega}{2}\right)4\pi\hbar\Omega\Im\left\{\frac{1}{\epsilon(\vec{k},\omega)} - 1\right\}.$$

Since $\delta\vec{\mathcal{E}}(0) = \delta\vec{\mathcal{E}}(0)^+$ it follows that for $\vec{k} \to 0$ the left hand side coincides with the fluctuation $\delta\vec{\mathcal{E}}(0)_\omega^2$ and thus

$$\delta\vec{\mathcal{E}}(0)_\omega^2 = -\frac{1}{2}\coth\left(\frac{\beta\hbar\omega}{2}\right)4\pi\hbar\Omega\Im\left\{\frac{1}{\epsilon(0,\omega)}\right\}.$$

Nevertheless,

$$\Im\left\{\frac{1}{\epsilon(0,\omega)} - 1\right\} = \frac{-\Im\epsilon(0,\omega)}{\Re\epsilon(0,\omega)^2 + \Im\epsilon(0,\omega)^2},$$

and since at $\omega \to 0$ only the imaginary part of the dielectric function diverges as $\frac{4\pi}{\omega}\sigma_{dc}$, we get in this limit

$$\lim_{\omega\to 0}\delta\vec{\mathcal{E}}(0)_\omega^2 = k_BT\frac{\Omega}{\sigma_{dc}}.$$

Now, in the case of a constant field $\vec{\mathcal{E}}$ in space, its Fourier transform is proportional to the volume Ω. Therefore, finally we get the Nyquist theorem

$$\langle \delta \mathcal{E}^2 \rangle_0 = \frac{k_B T}{\Omega \sigma_{dc}} .$$

A more general statement emerges [Bányai, Gartner and Aldea (1984)] also, namely, a generalized Nyquist theorem for the noise of the $\vec{k} = 0$ component of the field at a finite frequency ω:

$$\delta \vec{\mathcal{E}}_\omega^2 = -\frac{1}{2\Omega} \coth\left(\frac{\beta \hbar \omega}{2}\right) 4\pi\hbar\Im\left\{\frac{1}{\epsilon(0,\omega)}\right\} . \tag{12.3}$$

As we have seen, the essential point in the derivation of the Nyquist theorem was the correct treatment of the electromagnetic potential, taking into account that it is a dynamical variable determined by the internal charge. This aspect was often ignored in early discussions of the linear response.

Chapter 13

The Kubo formulae

Let us now consider in more detail the linear response to an electric field, while ignoring the Coulomb interaction between the carriers. As we already mentioned, this corresponds to the Hartree (mean-field) approximation. The so called Kubo formula [Kubo (1957)] for the electric conductivity, as well as the formulae for the non-mechanical kinetic coefficients [Mori (1956); Kubo, Yokota and Nakajima (1957); Luttinger (1964)] are formulated within this frame. We will point out that their validity is also restricted by the validity of this approximation. Nevertheless, these formulae have a wide domain of phenomenological applications.

13.1 The Kubo formula for the electric conductivity

We start again from the linear response formula Eq. (11.2), which we shall bring to another form with the help of the Kubo identity

$$\left[A, e^{-\beta H}\right] = e^{-\beta H} \int_0^\beta d\lambda e^{\lambda H} \left[H, A\right] e^{-\lambda H} .$$

This operatorial identity for an arbitrary operator A is easy to check by taking its matrix element between eigenstates of the Hamiltonian H. Then we may write for the commutator of the perturbation H' with the equilibrium density operator ρ_0

$$\left[H', \rho_0\right] = \rho_0 \int_0^\beta d\lambda \left[H, H'_H(-\imath\hbar\lambda)\right] ,$$

and we may rewrite accordingly the first line of Eq. (11.2) as

$$\langle A(t)\rangle - \langle A\rangle_0 = \frac{1}{\imath\hbar}\int_{-\infty}^{t}dt'\int_0^{\beta}d\lambda\langle[H,H'_H(-\imath\hbar\lambda)]\,A_H(t-t')\rangle_0\,.$$

Here we have already put $t_0 \to -\infty$. Now, let us consider as perturbation the linear terms of the coupling to the (time dependent) electromagnetic potentials

$$H' = \int d\vec{x}\left\{\rho(\vec{x})V(\vec{x},t) - \vec{j}(\vec{x})\vec{A}(\vec{x},t)\right\}\,.$$

Here the $\vec{A}^2$ term of the non-relativistic theory was ignored, since it is of second order in the field. Instead of the charge density $\rho(\vec{x})$ of the previous discussions we consider now the average of the current density $\vec{j}(\vec{x})$:

$$\langle\vec{j}(\vec{x},t)\rangle = -\int_{-\infty}^{t}dt'\int_0^{\beta}d\lambda\int d\vec{x}'\langle\Big(\dot{\rho}_H(\vec{x}',-\imath\hbar\lambda)V(\vec{x}',t)$$
$$-\dot{\vec{j}}_H(\vec{x}',-\imath\hbar\lambda)\vec{A}(\vec{x}',t)\Big)\vec{j}_H(\vec{x},t-t')\rangle_0\,.$$

Using the continuity equation

$$\nabla\vec{j} + \frac{\partial}{\partial t}\rho = 0\,,$$

and after a partial integration in time, respectively coordinate, one recovers the electromagnetic field $\vec{E}$ getting (in vector components with $\mu,\nu = 1,2,3$, by omitting the index H indicating the Heisenberg picture for operators)

$$\langle j_\mu(\vec{x},t)\rangle = \sum_{\nu=1}^{3}\int_{-\infty}^{t}dt'\int_0^{\beta}d\lambda\int d\vec{x}'\langle j_\nu(\vec{x}',-\imath\hbar\lambda)j_\mu(\vec{x},t-t')\rangle_0 E_\nu(\vec{x}',t')\,.$$

Again, we have a nonlocal relationship in time and space between the average current density and the electromagnetic field.

Since we have a convolution in the time variable, a Fourier transform simplifies the formula. Even more, since in a homogeneous medium the correlation functions may depend only on the difference of the coordinates, a Fourier transformation here is also appropriate. Therefore, one may define a frequency and wave-vector dependent conductivity tensor

$$\sigma_{\mu\nu}(\omega,\vec{k}) = \int_{-\infty}^{0}dt\int_0^{\beta}d\lambda\int d\vec{x}e^{\imath(\vec{k}\vec{x}+\omega t)}\langle j_\nu(\vec{x},t-\imath\hbar\lambda)j_\mu(0,0)\rangle_0\,.$$

In the following, we will concentrate on the $\vec{k} = 0$ conductivity $\sigma_{\mu\nu}(\omega, 0)$, which we denote by $\sigma_{\mu\nu}(\omega)$. In this case of a homogeneous field in a homogeneous medium, one may introduce an "overall current density" operator $j_\mu \equiv \frac{1}{\Omega} \int d\vec{x} j_\mu(\vec{x})$ and the Kubo formula [Kubo (1957)] looks as

$$\sigma_{\mu\nu}(\omega) = \Omega \int_{-\infty}^{0} dt e^{\imath\omega t} \int_{0}^{\beta} d\lambda \langle j_\nu(t - \imath\hbar\lambda) j_\mu(0) \rangle_0 \, .$$

Actually, one has to take into account that in order to extend the integration in time over an infinite interval, one has to consider an adiabatic introduction of the field, which can be removed only after the thermodynamic limit. Therefore, the proper Kubo formula for the electric conductivity looks like

$$\sigma_{\mu\nu}(\omega) = \lim_{s \to +0} \lim_{\Omega \to \infty} \Omega \int_{-\infty}^{0} dt e^{\imath\omega t} e^{st} \int_{0}^{\beta} d\lambda \langle j_\nu(t - \imath\hbar\lambda) j_\mu(0) \rangle_0 \, , \quad (13.1)$$

and the infinite volume limit is to be understood in the thermodynamic sense, i.e. at a fixed average carrier density.

Another version of the Kubo formula may be obtained by expressing the "overall current density" operator j_μ as the time derivative of the "overall coordinate" operator $X_\mu = \int d\vec{x} x_\mu n(\vec{x})$

$$j_\mu = \frac{e}{\Omega} \frac{d}{dt} X_\mu = -\frac{e}{\Omega \imath\hbar} [H, X_\mu] \, . \quad (13.2)$$

This relationship is based just on the use of the continuity equation. Then by taking into account that

$$\langle j_\nu(t - \imath\hbar\lambda) j_\mu(0) \rangle_0 = \frac{e^2}{\Omega^2 \hbar^2} \frac{d^2}{dt^2} \langle X_\nu(t - \imath\hbar\lambda) X_\mu(0) \rangle_0 \, ,$$

integrating by parts the Kubo formula Eq. (13.1) in the time variable, and using the identity

$$\frac{d}{dt} \int_{0}^{\beta} d\lambda \langle X_\nu(t - \imath\hbar\lambda) X_\mu(0) \rangle_0 = \frac{1}{\imath\hbar} \langle [X_\mu(0), X_\nu(t)] \rangle_0 \, ,$$

one gets the coordinate version of the Kubo formula

$$\sigma_{\mu\nu}(\omega) = \lim_{s\to+0}\lim_{\Omega\to\infty}\frac{e^2\omega^2}{\Omega}\int_{-\infty}^0 dt e^{i\omega t}e^{st}\int_0^\beta d\lambda$$

$$\times\langle X_\nu(-i\hbar\lambda)\,(X_\mu(-t) - X_\mu(0))\rangle_0\,,$$

or

$$\sigma_{\mu\nu}(\omega) = -\lim_{s\to+0}\lim_{\Omega\to\infty}\frac{e^2\omega}{\Omega\hbar}\int_{-\infty}^0 dt e^{i\omega t}e^{st}\langle[X_\mu(0), X_\nu(t)]\rangle_0\,. \qquad (13.3)$$

The diagonal part of the static conductivity tensor can be brought further to a form which shows the close relationship with the diffusion problem

$$\sigma_{\mu\mu}(0) = \lim_{t\to\infty}\lim_{\Omega\to\infty}\frac{e^2}{2\Omega t}\int_0^\beta d\lambda\langle(X_\mu(t) - X(0))\,(X_\mu(t + i\hbar\lambda) - X_\mu(i\hbar\lambda))\rangle_0.$$

13.2   The non-mechanical kinetic coefficients

Until now we discussed the problem of the kinetic coefficients (susceptibilities, such as the conductivity) that correspond to the reaction of a system to external forces present in a perturbative term in the Hamiltonian. Another important phenomenological problem is the reaction to non-mechanical "forces", like a gradient of concentration or temperature. The derivation of the corresponding response formulae [Mori (1956); Kubo, Yokota and Nakajima (1957)] followed in time closely that of the Kubo formula for the electric conductivity. Here we are not going to give the original derivations (see also [Kubo, Toda and Hashitsume (1985)]), but a later ingenious solution [Luttinger (1964)] to this problem, formulated as a generalization of the Einstein relation.

Let us define the stationary mechanical and non-mechanical kinetic coefficients $L_{\mu\nu}^{(i)}$ and $\tilde{L}_{\mu\nu}^{(i)}(i = 1, 2, 3, 4)$ by the phenomenological relationships for the average particle and energy currents in a homogeneous medium

$$\langle j_\mu\rangle = L_{\mu\nu}^{(1)}\partial_\nu V - \tilde{L}_{\mu\nu}^{(1)}T\partial_\nu\left(\frac{\mu}{T}\right) - \tilde{L}_{\mu\nu}^{(2)}T\partial_\nu\left(\frac{1}{T}\right) + L_{\mu\nu}^{(2)}\partial_\nu\psi \quad (13.4)$$

$$\langle j_\mu^E\rangle = L_{\mu\nu}^{(3)}\partial_\nu V - \tilde{L}_{\mu\nu}^{(3)}T\partial_\nu\left(\frac{\mu}{T}\right) - \tilde{L}_{\mu\nu}^{(4)}T\partial_\nu\left(\frac{1}{T}\right) + L_{\mu\nu}^{(4)}\partial_\nu\psi\,.$$

Here T is the temperature, μ is the chemical potential, while $V(\vec{x})$ is an electric potential (energy), and $\psi(\vec{x})$ a fictious gravitational potential coupled to the energy density operator $h(\vec{x})$. The mechanical coefficients $L_{\mu\nu}^{(i)}$ may be calculated with the linear response theory with regard to the perturbation

$$F = \int d\vec{x} \left\{ n(\vec{x})V(\vec{x}) + h(\vec{x})\psi(\vec{x}) \right\} ,$$

and by a slight generalization of the (stationary) Kubo formula one gets

$$L_{\mu\nu}^{(1)} = \lim_{s\to+0} \lim_{\Omega\to\infty} \Omega \sum_{\nu=1}^{3} \int_{-\infty}^{0} dt e^{st} \int_{0}^{\beta} d\lambda \langle j_\nu(t - \imath\hbar\lambda) j_\mu(0) \rangle_0 , \quad (13.5)$$

$$L_{\mu\nu}^{(2)} = \lim_{s\to+0} \lim_{\Omega\to\infty} \Omega \sum_{\nu=1}^{3} \int_{-\infty}^{0} dt e^{st} \int_{0}^{\beta} d\lambda \langle j_\nu^E(t - \imath\hbar\lambda) j_\mu(0) \rangle_0 ,$$

$$L_{\mu\nu}^{(3)} = \lim_{s\to+0} \lim_{\Omega\to\infty} \Omega \sum_{\nu=1}^{3} \int_{-\infty}^{0} dt e^{st} \int_{0}^{\beta} d\lambda \langle j_\nu(t - \imath\hbar\lambda) j_\mu^E(0) \rangle_0 ,$$

$$L_{\mu\nu}^{(4)} = \lim_{s\to+0} \lim_{\Omega\to\infty} \Omega \sum_{\nu=1}^{3} \int_{-\infty}^{0} dt e^{st} \int_{0}^{\beta} d\lambda \langle j_\nu^E(t - \imath\hbar\lambda) j_\mu^E(0) \rangle_0 .$$

The energy current density operator is supposed to satisfy the continuity equation with the energy density operator

$$\dot{h}(\vec{x}) + \partial_\mu j_\mu^E(\vec{x}) = 0 ,$$

in analogy to the usual continuity equation for the particle current and density. However, one must emphasize here that the existence of a local energy density operator is not at all obvious and does not include the case of long-range Coulomb interactions. Again, we meet the same limitation as by the discussion of the Kubo formula itself.

We will show in what follows that $\tilde{L}_{\mu\nu}^{(i)} = L_{\mu\nu}^{(i)}$. The proof is based on the idea that in the presence of all these forces an equilibrium state is possible, in which the total currents vanish. Let us consider such a situation, in which the system is in equilibrium in the presence of the potentials V and ψ. Of course, these potentials produce particle-density and energy-density gradients, which on their turn may be interpreted as gradients of the chemical potential and of the temperature by:

$$\partial_\nu \left(\frac{\mu}{T} \right) = \frac{\partial \left(\frac{\mu}{T} \right)}{\partial \langle n \rangle} \partial_\nu \langle n \rangle + \frac{\partial \left(\frac{\mu}{T} \right)}{\partial \langle h \rangle} \partial_\nu \langle h \rangle \qquad (13.6)$$

$$\partial_\nu \left(\frac{1}{T} \right) = \frac{\partial \left(\frac{1}{T} \right)}{\partial \langle n \rangle} \partial_\nu \langle n \rangle + \frac{\partial \left(\frac{1}{T} \right)}{\partial \langle h \rangle} \partial_\nu \langle h \rangle .$$

In equilibrium

$$\langle n(\vec{x}) \rangle = \frac{1}{Tr \left\{ e^{-\beta(H+F-\mu' N)} \right\}} Tr \left\{ e^{-\beta(H+F-\mu' N)} n(\vec{x}) \right\} ,$$

where μ' is the chemical potential in the presence of the perturbation F, as contrasted to the chemical potential μ of the system in equilibrium without the external potentials. However, this distinction plays no role in a first order theory. Retaining the linear terms of the expansion

$$e^{-\beta(H+F-\mu' N)} = e^{-\beta(H-\mu' N)} \left\{ 1 - \int_0^\beta d\lambda F(-\imath \hbar \lambda) + \cdots \right\} ,$$

we get

$$\langle n(\vec{x}) \rangle = \langle n(\vec{x}) \rangle_0 - \int_0^\beta d\lambda \int d\vec{x}' (\langle n(\vec{x}', -\imath \hbar \lambda) n(\vec{x}) \rangle_0$$
$$- \langle n(\vec{x}', -\imath \hbar \lambda) \rangle_0 \langle n(\vec{x}) \rangle_0) e V(\vec{x}') + \cdots .$$

Here, for sake of simplicity, we considered only an electromagnetic potential $V(\vec{x})$ and the averaging in $\langle ... \rangle_0$ is understood over the equilibrium of the unperturbed system. Taking into account also the homogeneity of the system in the absence of the fields, that implies a coordinate independent $n(\vec{x})\rangle_0$ and a correlation $\langle n(\vec{x}', -\imath \hbar \lambda) n(\vec{x}) \rangle_0$ depending only on the difference of the coordinates, we have

$$\partial_\mu \langle n(\vec{x}) \rangle = - \int_0^\beta d\lambda \int d\vec{x}' (\langle n(\vec{x}', -\imath \hbar \lambda) n(\vec{x}) \rangle_0$$
$$- \langle n(0, -\imath \hbar \lambda) \rangle_0 \langle n(0) \rangle_0) e \partial_\mu V(\vec{x}') .$$

Further, in a potential with a constant gradient (homogeneous field) we have

$$\int d\vec{x}' (\langle n(\vec{x}', -\imath\hbar\lambda)n(\vec{x})\rangle_0 - \langle n(0, -\imath\hbar\lambda)\rangle_0\langle n(0)\rangle_0) = \frac{1}{\Omega}\left(\langle N^2\rangle - \langle N\rangle^2\right),$$

and we get

$$\partial_\mu\langle n\rangle = -\beta\frac{1}{\Omega}\left(\langle N^2\rangle - \langle N\rangle^2\right)e\partial_\mu V = \frac{\partial\langle N\rangle}{\partial\beta\mu},$$

and analogously

$$\partial_\mu\langle h\rangle = -\beta\frac{1}{\Omega}\left(\langle HN\rangle - \langle H\rangle\langle N\rangle\right)e\partial_\mu V = \frac{\partial\langle H\rangle}{\partial\beta\mu}.$$

Inserting into Eqs.(13.6) one gets

$$\partial_\nu\left(\frac{\mu}{T}\right) = -\frac{1}{T}e\partial_\mu V$$

and

$$\partial_\nu\left(\frac{1}{T}\right) = 0.$$

Therefore, from the vanishing of the average currents in Eqs.(13.4) one gets

$$\tilde{L}_{\mu\nu}^{(1)} = L_{\mu\nu}^{(1)}, \tag{13.7}$$

and

$$\tilde{L}_{\mu\nu}^{(3)} = L_{\mu\nu}^{(3)}. \tag{13.8}$$

Similarly, in equilibrium in the presence of a gravitational field one gets

$$\partial_\nu\left(\frac{\mu}{T}\right) = 0$$

and

$$\partial_\nu\left(\frac{1}{T}\right) = -\frac{1}{T}\partial_\nu\psi.$$

Then again from the condition of vanishing of the average currents one gets also

$$\tilde{L}^{(2)}_{\mu\nu} = L^{(2)}_{\mu\nu} \tag{13.9}$$

and

$$\tilde{L}^{(4)}_{\mu\nu} = L^{(4)}_{\mu\nu} \, . \tag{13.10}$$

While Eq.(13.7) is just the well-known Einstein relation Eq.(10.1) between the conductivity and diffusivity, Eqs.(13.8)-(13.10) are its generalizations. Indeed, from the identifications $L^{(1)}_{\mu\nu} \equiv \frac{1}{e^2}\sigma_{\mu\nu}$ and $\tilde{L}^{(1)}_{\mu\nu} \equiv D_{\mu\nu}/\frac{\partial n}{\partial \mu}$, follows Eq.(10.1).

13.3 Symmetry relations, dispersion relations and sum rules

Let us discuss now some of the general properties of the Kubo formula Eq. (13.1). In order to simplify the notations, the $s \to +0$ limit, followed by the thermodynamic one, although implicitly understood, will be omitted in what follows. Introducing into the Kubo formula the complete set of eigenstates $|n\rangle$ of the Hamiltonian H one gets

$$\sigma_{\mu\nu}(\omega) = \frac{\Omega}{Z} \int dt e^{\imath\omega t} e^{st} \int_0^\beta d\lambda$$
$$\times \sum_{m,n} e^{-\beta(E_m - \mu N_m)} e^{\frac{1}{\hbar}(E_n - E_m)(t + \imath\hbar\lambda)} \langle m|j_\nu|n\rangle\langle n|j_\mu|m\rangle \, .$$

Performing the integrations and using the well known relation

$$\lim_{s \to +0} \frac{1}{x \pm \imath s} = P\frac{1}{x} \mp \imath\pi\delta(x)$$

from the theory of distributions, one may put the previous equation under the form

$$\sigma_{\mu\nu}(\omega) = -\hbar\Omega \sum_{m,n} \frac{\rho_0(E_n) - \rho_0(E_m)}{E_n - E_m} \langle m|j_\nu|n\rangle\langle n|j_\mu|m\rangle$$
$$\times \left(\pi\delta(E_n - E_m - \hbar\omega) - \imath P\frac{1}{E_n - E_m - \hbar\omega} \right) ,$$

where $\rho_0(E_n) \equiv \frac{1}{Z} e^{-\beta(E_m - \mu N_m)}$. Under this form one can see immediately the following symmetry relations

$$\Re \sigma_{\mu\nu}(\omega) = \Re \sigma_{\mu\nu}(-\omega); \quad \Im \sigma_{\mu\nu}(\omega) = \Im \sigma_{\mu\nu}(-\omega). \tag{13.11}$$

If one takes into account also that $\langle n | j_\mu | m \rangle = \langle m | j_\mu | n \rangle^* = -\langle \tilde{m} | j_\mu | \tilde{n} \rangle$, where the state $|\tilde{n}\rangle$ belongs to the same eigenenergy E_n of the problem with the reversed magnetic field $\vec{B}$, one gets also

$$\sigma_{\mu\nu}(\omega; \vec{B}) = \sigma_{\nu\mu}(\omega; \vec{B}). \tag{13.12}$$

Another important consequence of the above Kubo formula is the dispersion relation

$$\Im \sigma_{\mu\nu}(\omega) = -P \int_{-\infty}^{\infty} d\omega' \frac{\Re \sigma_{\mu\nu}(\omega')}{\omega' - \omega}. \tag{13.13}$$

Although it follows formally from Eq. (13.11), its validity is not at all guaranteed, being conditioned by the convergence of the integral. Eventually, some subtractions are necessary, as in the case of the related Kramers-Kroenig relation. Another problem with this relationship is that it extends over a frequency domain that is far beyond the validity of the non-relativistic quantum mechanical models of condensed matter. Nevertheless, such relationships are successfully used in the phenomenology.

Eq. (13.11) allows also to write down a nice sum rule. Let us consider the integral

$$\int_{-\infty}^{\infty} d\omega \Re \sigma_{\mu\nu}(\omega) = -\pi\Omega \sum_{m,n} \frac{\rho_0(E_n) - \rho_0(E_m)}{E_n - E_m} \langle m | j_\nu | n \rangle \langle n | j_\mu | m \rangle.$$

Expressing just one of the currents by the time derivative of the "overall coordinate" Eq. (13.2), one may get rid of the energy denominator and remains with

$$\frac{\imath\pi e}{\hbar} \sum_n (\rho_0(E_n) - \rho_0(E_m)) \langle m | j_\nu | n \rangle \langle n | X_\mu | m \rangle = \frac{\imath\pi e}{\hbar} Tr \{ \rho_0 [j_\nu, X_\mu] \}.$$

On the other hand, the commutation relationship between the current density and the "overall coordinate" operator is related to the particle number operator by $[j_\mu, X_\nu] = -\frac{ie\hbar}{m}\frac{N}{\Omega}\delta_{\mu\nu}$, and therefore we get

$$\int_{-\infty}^{\infty} d\omega \Re\sigma_{\mu\nu}(\omega) = \pi\frac{e^2}{m}\frac{\langle N\rangle}{\Omega}\delta_{\mu\nu}. \qquad (13.14)$$

Again, one must remark that these algebraic manipulations are highly questionable, not only because the formerly mentioned problems (convergence of integrals and extension of the non-relativistic theory over its domain of validity), but also because of the mathematically strictly forbidden manipulation of the velocity and coordinate operators under the trace. (Such manipulations may lead to absurdities, since these two operators have neither the same domain nor the same range in the Hilbert space.). Nevertheless, this sum rule has proven itself useful in phenomenological applications.

On the basis of the above formulae one can derive also a useful representation for the real part of the conductivity, connecting it to the imaginary part of the retarded Green function of the currents:

$$\Re\sigma_{\mu\nu}(\omega) = -\pi\hbar\Omega \sum_{m,n} \frac{\rho_0(E_n) - \rho_0(E_m)}{E_n - E_m}\langle m|j_\nu|n\rangle\langle n|j_\mu|m\rangle\delta(E_n - E_m - \hbar\omega)$$

$$= \frac{\Omega}{2\hbar\omega}(e^{\beta\hbar\omega} - 1)\int_{-\infty}^{\infty} dt\langle j_\mu(t)j_\nu(0)\rangle e^{-i\omega t} \qquad (13.15)$$

$$\equiv \frac{2\pi}{\hbar\omega}\Im G_r^{\mu\nu}(\omega),$$

hence, eliminating the somewhat disturbing imaginary time integration.

13.4   Implementation of the Master (rate) — approach into the Kubo formulae. Transport by scattered electrons versus transport by hopping electrons

The convergence of the infinite time integration in the Kubo formula is intimately related to irreversibility by the decay of the correlations. The relaxation due to weak interaction with a thermal bath, like scattering on equilibrium phonons, may be implemented within the Master or rate equation approximation [Bányai and Aldea (1979)] which - as we will see later - can be justified by the van Hove limit. We will discuss this case here.

If we are interested either in the problem of weakly scattered free electrons (on phonons) within the frame of the current-correlator version of the Kubo formula Eq. (13.1) or in the problem of weakly phonon assisted hopping of localized electrons within the coordinate-correlator version of the Kubo formula Eq.(13.3), we have to calculate correlators of the type

$$\phi_{ij}(t) \equiv \frac{1}{\beta} \int_0^\beta d\lambda \langle n_i n_j(t + i\hbar\lambda) \rangle_0 .$$

Here the indices i, j of the occupation number operators n_i, n_j run either over the plane wave states of wave vector $\vec{k}$ (for free electrons) or over the discrete coordinates $\vec{R}$ of the localized states. In both cases the strength of the interaction with the thermal bath is specified by a coupling constant g. Under the assumption that this coupling is sufficiently weak, one might neglect it everywhere, except the real time t dependence, i.e.

$$\phi_{ij}(t) \approx \langle n_i n_j(t) \rangle_0^0 ,$$

where the upper index ^{0} indicates averaging over the independent macro-canonical distributions of the system and of the thermostat. On the other hand,

$$\langle n_i n_j(t) \rangle_0^0 = \sum_{n,n'} P_n^0 P_{n'}^{(n)}(t) n_i n_j' ,$$

where $P_n^0 \equiv \frac{1}{Z} e^{-\beta(\sum(\epsilon_i - \mu)n_i)}$, and

$$P_{n'}^{(n)}(t) \equiv \sum_{\nu,\nu'} P_\nu^0 \left| \left(e^{-\frac{i}{\hbar} Ht} \right)_{n,\nu:n',\nu'} \right|^2 ,$$

is the probability to find the electrons in the state n' at the momentum t, independently of the state of the phonons, if at $t = 0$ they were in the state n, while the phonons were in thermal equilibrium described by P_ν^0. In the Master equation approximation of Eq. (8.4) for $P_{n'}^{(n)}(t)$ one gets

$$\phi_{ij}(t) \approx \sum_{n,n'} P_n^0 \left(e^{-At} \right)_{n'n} n_i n_j' ,$$

with the evolution operator A of the Master equation.

Further, using the Master-Boltzmann approximation Eq. (8.9), one gets

$$\phi_{ij}(t) \approx f_i(1 - f_j) \left(e^{-\Gamma t} \right)_{ij} + f_i f_j , \qquad (13.16)$$

where Γ is the evolution operator of the linearized rate equation.

Since for low frequencies the time integral in the Kubo formulae is dominated by the long-time behavior of the correlator, the use of this Markovian approximation seems to be justified, and we are going to implement it.

1. First let us consider the case of free electrons scattered on a thermal bath (phonons). The "overall current" operator here is diagonal in the basis of the unperturbed Hamiltonian

$$H_0 = \sum_{\vec{k}} \varepsilon_k n_{\vec{k}}$$

with $\varepsilon_k \equiv \frac{\hbar^2 k^2}{2m}$, and is given by

$$j_\mu = \frac{e\hbar}{m} \sum_{\vec{k}} k_\mu n_{\vec{k}} .$$

Using now the Master-Boltzmann approximation Eq. (8.9) for the correlator of the $\vec{k}$ -state occupation numbers in the Kubo formula Eq. (13.1), one gets

$$\sigma_{\mu\nu}(\omega)^{MB} = -\frac{e^2 \hbar^2}{m^2 \Omega} \sum_{\vec{k},\vec{k}'} \frac{df_k}{d\varepsilon_k} k_\mu \left(\frac{1}{\imath\omega + \Gamma} \right)_{\vec{k},\vec{k}'} k'_\nu . \qquad (13.17)$$

Here a matrix notation was used.

If one ignores here the energy-loss to the thermostat, which is justified in the case of scattering on acoustical phonons, then $|\vec{k}|^2 = |\vec{k}'|^2$ and using also the rotational invariance, one can define a scalar relaxation time by

$$\sum_{\vec{k}'} \Gamma_{\vec{k},\vec{k}'} k'_\mu = k_\mu \tau^{-1}(|\vec{k}|) .$$

As it may be seen by its successive application in the series development in Γ of Eq. (13.17), it gives rise to the standard textbook result

$$\sigma_{\mu\nu}(\omega)^{MB} = -\frac{4\pi e^2 \hbar^2}{3m^2 \Omega} \int_0^\infty dk k^2 \frac{df_k}{d\varepsilon_k} \frac{1}{\imath\omega + \tau(k)^{-1}}$$

which usually is derived from the Boltzmann equation approach.

2. Another important case is the dual model of conduction of localized electrons hopping from site to site due to emission and absorption of phonons. In this case the "overall coordinate" operator $\vec{X}$ defined by

$$X_\mu = \sum_{\vec{R}} R_\mu n_{\vec{R}}$$

commutes with the unperturbed Hamiltonian

$$H_0 = \sum_{\vec{R}} \varepsilon_{\vec{R}} n_{\vec{R}},$$

where $n_{\vec{R}}$ are the occupation number operators of the site with coordinate $\vec{R}$.

Now, we have to use the the Master-Boltzmann approximation in the coordinate version of the Kubo formula Eq. (13.3), and one gets, again, in a matrix notation

$$\sigma_{\mu\nu}(\omega)^{MB} = -\frac{e^2}{\Omega} \sum_{\vec{R},\vec{R}'} \frac{df(\varepsilon_{\vec{R}})}{d\varepsilon_{\vec{R}}} R_\mu \left(\frac{\imath\omega\Gamma}{\imath\omega + \Gamma} \right)_{\vec{R},\vec{R}'} R_\nu. \qquad (13.18)$$

Due to the positivity of the matrix Γ, the real part of the conductivity tensor

$$\Re\sigma_{\mu\nu}(\omega)^{MB} = -\frac{e^2}{\Omega} \sum_{\vec{R},\vec{R}'} \frac{df(\varepsilon_{\vec{R}})}{d\varepsilon_{\vec{R}}} R_\mu \left(\frac{\omega^2\Gamma}{\omega^2 + \Gamma^2} \right)_{\vec{R},\vec{R}'} R_\nu$$

is a positive, monotonously increasing function of the frequency ω.

A peculiarity of this formula is that at a first glance it looks as it would imply a vanishing d.c. conductivity. Yet this is not true. The mathematically delicate point is that, according to the definitions, one has first to perform the infinite volume limit ($\Omega \to \infty$), and only after that the zero frequency limit ($\omega \to 0$), while the two operation do not commute. In order to illustrate this important point, let us consider that the localized states have identical energies ε, and are distributed on a cubic Bravais lattice of lattice constant a (also the infinite volume limit is already assumed), with translations invariant transition rates $w(\vec{R}, \vec{R}') \equiv w(\vec{R} - \vec{R}')$, which are assumed to decrease rapidly with the distance, as in the example we discussed earlier in Chapter 10.3. In this case, the matrix $\Gamma_{\vec{R},\vec{R}'} = \Gamma(\vec{R} - \vec{R}')$ is translation invariant. As we have discussed, an important property of this matrix is that it is positive, and due to the connectivity of the lattice has a unique zero eigenvalue corresponding to the eigenvectors with equal components $\sum_{\vec{R}} \Gamma(\vec{R}) = 0$. Using these properties, one may bring Eq. (13.18) to the form

$$\sigma_{\mu\nu}(\omega)^{MB} = \frac{e^2}{a^3} \frac{df(\varepsilon)}{d\varepsilon} \sum_{\vec{R}} R_\mu R_\nu \left(\frac{\imath\omega\Gamma(\vec{R})}{\imath\omega + \Gamma(\vec{R})} \right),$$

or after a Fourier transformation

$$\sigma_{\mu\nu}(\omega)^{MB} = -\frac{e^2}{a^3} \frac{df(\varepsilon)}{d\varepsilon} \frac{\partial^2}{\partial k_\mu \partial k_\nu} \left(\frac{\imath\omega\tilde{\Gamma}(\vec{k})}{\imath\omega + \tilde{\Gamma}(\vec{k})} \right)\Bigg|_{\vec{k}=0}.$$

Now, since $\tilde{\Gamma}(0) = 0$ and the transition rates are supposed to be symmetrical, the low $\vec{k}$ development of $\tilde{\Gamma}(\vec{k})$ starts with a quadratic term - as was earlier discussed - i.e. $\tilde{\Gamma}(\vec{k}) \approx D\vec{k}^2$, with the diffusion constant $D = \frac{1}{6} \sum_{\vec{R}} w(\vec{R})\vec{R}^2$. Therefore, we get

$$\sigma_{\mu\nu}(\omega)^{MB} = -\delta_{\mu\nu} \frac{e^2}{a^3} \frac{df(\varepsilon)}{d\varepsilon} D,$$

i.e. a frequency independent, diagonal, real and finite conductivity tensor, as one might have expected on the basis of the formerly discussed diffusive behavior of the hopping model.

The model of electrons hopping between sites is appropriate for the description of low temperature transport on localized states in disordered materials, but also for the peculiar problem of the transverse magneto-resistance in a strong magnetic field, we discuss in the next Section.

An important ingredient in obtaining these results was the approximation that the interaction - which has a decisive role in the time evolution - may be ignored in the definition of the initial equilibrium distribution, i.e. one ignores the initial correlations.

13.5   Ideal relaxation

We are going to illustrate the signification of the Kubo formula on the example of an assumed ideal relaxation. If one introduces a supplementary artificial relaxation term [Lax (1958]

$$-i\frac{\rho - \rho_0}{\tau}$$

in the Liouville equation of the density operator ρ, then one does not need the adiabatic introduction of the external perturbation, and the inverse relaxation time $\frac{1}{\tau}$ takes its place in the resulting Kubo formula

$$\sigma_{\mu\nu}(\omega) = \lim_{\Omega\to\infty} \Omega \int_{-\infty}^{0} dt e^{i\omega t} e^{\frac{1}{\tau}t} \int_{0}^{\beta} d\lambda \langle j_\nu(t - i\hbar\lambda) j_\mu(0)\rangle_0 \,.$$

Although the ideal relaxation term violates the gauge invariance of the theory, and therefore must be looked up with some suspicion, at least in the linear response, it creates no problems. Indeed, the above result is gauge independent and may be obtained also directly from the Kubo formula by just saying that the correlations (due to some unspecified interaction) decay exponentially. Then the current-current correlator here is that of the "unperturbed" theory, i.e. it does not take into account dissipation. Now one may consider the problem of free electrons of mass m in the presence of a magnetic field $\vec{B}$ along the z axis, for example in the Landau gauge. The current-current correlator of the free particles may be easily calculated in this case and one gets explicitely [Bányai (1968)]

$$\sigma_{zz}(\omega) = \frac{ne^2}{m} \frac{1}{i\omega + \tau^{-1}} \,,$$

$$\sigma_{xx}(\omega) = \frac{ne^2}{2m} \left(\frac{1}{i(\omega - \omega_c) + \tau^{-1}} + \frac{1}{i(\omega + \omega_c) + \tau^{-1}} \right), \quad (13.19)$$

$$\sigma_{xy}(\omega) = \frac{ne^2}{2m} \left(\frac{1}{i(\omega - \omega_c) + \tau^{-1}} - \frac{1}{i(\omega + \omega_c) + \tau^{-1}} \right),$$

$$\sigma_{xx}(\omega) = \sigma_{yy}(\omega)\,,$$

$$\sigma_{xy}(\omega) = -\sigma_{yx}(\omega)\,.$$

Here $\omega_c = \frac{e|B|}{mc}$ is the cyclotron frequency. The same result may be obtained also form the classical Boltzmann equation with a collision term describing an ideal relaxation.

One may remark the resemblance to the formula of ideal relaxation (in the absence of a magnetic field) Eq. (13.19) with the Master-Boltzmann result Eq.(13.17). There, however, one had to integrate over a whole spectrum of relaxation times.

On these expressions one may get also an idea about the significance of the initial correlations. Let us consider for example the static transverse magnetic conductivity as given by the above formulae

$$\sigma_{xx} = \frac{ne^2}{m} \frac{1}{1 + (\omega_c \tau)^2} \; .$$

Interpreting the phenomenological relaxation time τ as the average life-time of an one-electron state given by the inverse total transition probability to any other state, due to the interaction with the coupling strength g (electron-phonon or electron-impurity coupling), it follows that in the weak coupling limit $\tau \sim \frac{1}{g^2}$ and consequently

$$\sigma_{xx} \sim \begin{cases} g^{-2} & for \quad \omega_c \tau \ll 1 \\ g^2 & for \quad \omega_c \tau \gg 1 \end{cases} \; .$$

In the weak coupling limit the contribution of the initial correlation is of order g^2, and the correction to the conductivity due to initial correlations is certainly insignificant for weak magnetic fields ($\omega_c \tau \ll 1$). For a strong field ($\omega_c \tau \gg 1$), however, the correction may be well significant. Indeed, as revealed by a more sophisticated analysis, the contribution of the initial correlations to the transverse magneto-conductivity is substantial if the magnetic energy level spacing $\hbar \omega_c$ is much bigger than the thermal energy $k_B T$ (case of quantizing magnetic fields) [Argyres (1960)].

13.6 The transverse strong-field magneto-resistance

Let us consider free electrons in a magnetic field $\vec{B} = (0, 0, B)$ interacting with a thermostat. (phonons) If the field is very strong, namely $\omega_c \gg \tau^{-1}$ and $\hbar \omega_c \gg k_B T$, one has to take into account its quantizing role.

We choose the Landau gauge for the vector potential $\vec{A}(\vec{r}) = (0, Bx, 0)$. The eigenfunctions for an electron in magnetic field are

$$\psi_{n,X,k_z}(\vec{r}) = \phi_n(x - \mathcal{X}) \frac{e^{\imath(k_z z + k_y y)}}{\sqrt{L_z L_y}} \; ,$$

where $\mathcal{X} = sign(e) \ell_B^2 k_y$ is the coordinate of the cyclotron motion along the x-axis and $\ell_B = \sqrt{\frac{\hbar c}{|e|B}}$ is the magnetic length. The wave functions describing the motion along this axis are the oscillator ones

$$\phi_n(x) = \frac{1}{\sqrt{\ell_B}} e^{-x^2/2\ell_B^2} H_n(x/\ell_B) \,,$$

and the energy is infinitely degenerate in the quantum number $\mathcal{X}$

$$\epsilon_{n,\mathcal{X},k_z} = \frac{\hbar^2 k_z^2}{2m} + \hbar\omega_c(n + \frac{1}{2}) \qquad (n = 0, 1, 2, ...) \,.$$

Here $\omega_c = \frac{|e|B}{mc}$ is the cyclotron frequency. To simplify the notations, we often denote the whole set of quantum numbers $n, \mathcal{X}, k_z$ simply by ν.

One would think on intuitive grounds that by very high magnetic fields, for the charge transport along the x-axis, only the motion of the center of cyclotron motion is relevant. [Kubo, Hasegawa and Hashitsume (1959)] Since this coordinate is a constant of the motion in the absence of any interaction - without scattering - the transverse d.c. conductivity is vanishing ($\sigma_{xx}(0) = 0$) (see also Eq. (13.19). One may use this circumstance to perform a simple perturbational treatment of the interaction with the thermostat. Let the interaction with the thermostat (phonons) be described by the Hamiltonian

$$H_{eph} = \sum_\nu U_{\nu\nu'} a_\nu^+ a_{\nu'} \,,$$

where $U_{\nu\nu'}$ is an operator in the phonon space and let the states of the phonon system be indexed by some quantum numbers s with the corresponding energies ϵ_s. The Heisenberg equation of motion of the "overall coordinate" operator $X = \sum_\nu \mathcal{X} a_\nu^+ a_\nu$ is

$$i\hbar\dot{X} = [H_{eph}, X] = -\sum_{\nu,\nu'} U_{\nu\nu'}(\mathcal{X} - \mathcal{X}')a_\nu^+ a_{\nu'} \,.$$

If we introduce this into the Kubo formula Eq.(13.15) for the d.c. conductivity with the "overall current density", with j_x replaced by $\frac{e}{\Omega}\dot{X}$, then we get

$$\sigma_{xx}(0) = \frac{e^2 \beta}{2\Omega} \int_{-\infty}^{\infty} dt \langle \dot{X}(t)\dot{X}(0)\rangle_0 \,,$$

and we may see that $\Re\sigma_{xx}(\omega)$ starts already with the second order of the interaction. One may renounce to further corrections due to scattering and consider the evolution in real time as well as the averaging only in the non-interacting system. Then we get

$$\sigma_{xx}(0) = \frac{e^2\pi\beta}{\Omega\hbar} \sum_{\nu,\nu'} \sum_{\bar{\nu},\bar{\nu}'} (\mathcal{X} - \mathcal{X}')(\bar{\mathcal{X}} - \bar{\mathcal{X}}')\langle a_\nu^+ a_{\nu'} a_{\bar{\nu}}^+ a_{\bar{\nu}'}\rangle_0^0$$

$$\times \sum_{s,s'} \frac{1}{z} e^{-\beta\epsilon_s} U_{\nu\nu'}^{ss'} U_{\bar{\nu}\bar{\nu}'}^{\bar{s}\bar{s}'} \delta(E_{\nu'} + \epsilon_{s'} - E_\nu - \epsilon_s).$$

Using the relation

$$\langle a_\nu^+ a_{\nu'} a_{\bar{\nu}}^+ a_{\bar{\nu}'}\rangle_0^0 = \delta_{\nu\nu'}\delta_{\bar{\nu}\bar{\nu}'} f(E_\nu)f(E_{\bar{\nu}}) + \delta_{\nu\bar{\nu}'}\delta_{\bar{\nu}\nu'} f(E_\nu)\left(1 - f(E_{\bar{\nu}})\right),$$

valid for non-interacting fermions, with $f(E)$ being the Fermi-function, one gets finally Titeica's formula [Titeica (1935)]

$$\sigma_{xx}(0) = \frac{\beta e^2}{2\Omega} \sum_{\nu,\nu'} W_{\nu\nu'} (\mathcal{X} - \mathcal{X}')^2, \tag{13.20}$$

where $W_{\nu\nu'}$ is the "golden rule" transition rate of electrons from the state ν to the state ν' due to the interaction with the phonons (averaged over the equilibrium phonon state)

$$W_{\nu\nu'} = 2\pi f(E_\nu)\left(1 - f(E_{\bar{\nu}})\right) \sum_{s,s'} \frac{1}{z} e^{-\beta\epsilon_s} \left|U_{\nu\nu'}^{ss'}\right|^2 \delta(E_{\nu'} + \epsilon_{s'} - E_\nu - \epsilon_s).$$

One can see that the d.c. conductivity we got is identical with the one predicted by the hopping model on a continuum of sites. (Compare with the results of the previous section on hopping in a cubic lattice).

For high magnetic fields the magneto-conductivity shows characteristic oscillations, periodic on the $\frac{1}{B}$ scale (Shubnikov - de Haas effect). The high-field condition is most easily fulfilled for charge carriers with small effective mass, i.e. by semimetals and semiconductors. In semiconductors the low temperature conductivity is determined by the electron-impurity scattering. However, for elastic scattering by impurities the transverse conductivity as given by the Titeica formula Eq.(13.20) diverges in the weak coupling limit (Born approximation). To overcome this flaw, higher order scattering processes (collisional damping or multiple scattering) have to be taken into account [Davydov and Pomeranchuk (1940)].

Titeica's formula can be generalized to a.c. electric fields and arbitrary magnetic field strengths [Götze and Hajdu (1978); Götze and Hajdu (1979)].

In this Section the high-field transverse magneto-resistance formula was derived, using the assumption that only the center of cyclotronic motion matters. As we have seen, since the time derivative of this cyclotron coordinate is already proportional to the phonon coupling there is no place in this derivation for other corrections of the same order in the interaction. On the other hand, Titeica's formula Eq.(13.20) may be derived without this assumption [Argyres (1960)], by just performing a systematic development in $\frac{1}{\omega_c}$ (actually , as already mentioned, in $\frac{1}{\omega_c \tau}$). In this later derivation, however, other second order corrections in the interaction constant appear. Their origin lies in the fact that the equilibrium distribution $\frac{1}{Z} e^{-\beta(H_0 + H_{ph} + H_{eph})}$ itself is affected by the interaction. This "initial correlations" play an essential role in this later derivation.

Chapter 14

Excitonic absorption in a semiconductor

Now, in order to illustrate the application of the linear response, let us consider a simple model of a semiconductor at $T = 0$, where light absorption implies the excitation of Coulomb interacting electron-hole pairs from the vacuum. We admit that both the valence and conduction bands have simple isotropic parabolic dispersions with maximum (minimum) at $\vec{k} = 0$. We consider that the valence band is p-type with threefold degeneracy, while the conduction band is nondegenerate s-type. This is a good model for many direct-band semiconductors, which are important for applications. However, we ignore the spin, which leads only to a degeneracy factor. The kinetic energies of the electrons and holes are thus

$$e_e(\vec{k}) = \frac{\hbar^2 k^2}{2m_e} + \frac{1}{2}E_g \; ; \; e_{h_\mu}(\vec{k}) = \frac{\hbar^2 k^2}{2m_h} + \frac{1}{2}E_g \, .$$

For reasons of convenience, the p-states here are not taken in the usual $m = \pm 1, 0$ basis, but in the vector basis x, y, z. Obviously, the minimal energy to create a noninteracting electron-hole pair out of vacuum (fully occupied valence band) is the band gap E_g. As we shall see, due to their Coulomb attraction, it is, however, reduced with the binding energy.

The interaction with an optical electromagnetic field $\vec{E}$ inducing transitions between the valence and conduction bands is not easy to describe by the original minimal electromagnetic coupling, since, in order to have a treatable model of the semiconductor, one must use simplifying approximations. These approximations, however, destroy the gauge invariance of the theory.

The accepted simple phenomenological model to describe such a coupling describing creation and annihilation of electron-hole pairs out of the electron-hole vacuum is

$$H_{em}(t) = d \sum_{\mu=1}^{3} \int d\vec{x} \psi_{e,\sigma}(\vec{x})^{+} \psi_{h_\mu,-\sigma}(\vec{x})^{+} \mathcal{E}_\mu(\vec{x},t) + h.c.,$$

where d is an inter-band dipole matrix element, $\psi_e(\vec{x})$ is the annihilation operator of an electron in the point $\vec{x}$, while $\psi_{h_\mu}(\vec{x})$ is the corresponding operator for a hole in the band μ. This is an explicitly gauge invariant coupling, since it is expressed directly in terms of the electrical field strength. In optics this field is considered to be the solution of the Maxwell equations and not just the external field. Therefore, it includes also the field produced by the particles. If one neglects the coordinate dependence of the field - i.e. one takes the long wave-length limit - then one may rewrite this optical interband coupling as

$$H_{em}(t) = e\vec{X}\vec{\mathcal{E}}(t),$$

where we introduced an inter-band dipole operator as

$$eX_\mu \equiv d \sum_{\mu=1}^{3} \int d\vec{x} \psi_e(\vec{x})\psi_{h_\mu}(\vec{x}) + h.c..$$

Accordingly, we start from the coordinate version Eq. (13.3) of the Kubo formula with this inter-band dipole operator

$$\sigma_{\mu\nu}(\omega) = -\lim_{s\to+0}\lim_{\Omega\to\infty} \frac{d^2\omega}{\hbar} \int_{-\infty}^{0} dt e^{\imath\omega t} e^{st} \int d\vec{x} \int d\vec{x}'$$
$$\times \langle 0| \left[\psi_e(\vec{x},0)\psi_{h_\mu}(\vec{x},0) + h.c., \psi_e(\vec{x}',t)\psi_{h_\nu}(\vec{x},t) + h.c. \right] |0\rangle.$$

Here we already took into account that at $T = 0$ the averaging is on the ground state (electron-hole vacuum). Due to the fact that one cannot annihilate electrons and holes in the vacuum state, we are left with

$$\sigma_{\mu\nu}(\omega) = -\lim_{s\to+0}\lim_{\Omega\to\infty} \frac{d^2\omega}{\hbar} \int_{-\infty}^{0} dt e^{\imath\omega t} e^{st} \int d\vec{x} \int d\vec{x}'$$
$$\times \langle 0| \psi_e(\vec{x},0)\psi_{h_\mu}(\vec{x},0)\psi_e(\vec{x}',t)^{+}\psi_{h_\nu}(\vec{x}',t)^{+}$$
$$- \psi_e(\vec{x}',t)\psi_{h_\nu}(\vec{x}',t)\psi_e(\vec{x},0)^{+}\psi_{h_\mu}(\vec{x},0)^{+} |0\rangle.$$

Farther, taking into account that the average depends only on the difference of the coordinates, one gets

$$\sigma_{\mu\nu}(\omega) = -\lim_{s\to+0}\lim_{\Omega\to\infty}\frac{d^2\omega}{\hbar}\int_{-\infty}^{0}dt e^{\imath\omega t}e^{st}\int d\vec{x}$$

$$\times \langle 0|\psi_e(0,0)\psi_{h_\mu}(0,0)\psi_e(\vec{x},t)^+\psi_{h_\nu}(\vec{x},t)^+$$

$$- \psi_{e,}(\vec{x},t)\psi_{h_\nu}(\vec{x},t)\psi_e(0,0)^+\psi_{h_\mu}(0,0)^+|0\rangle. \tag{14.1}$$

Now, let us insert intermediate eigenstates of the Hamiltonian of the Coulomb interacting electrons and holes between the groups of annihilation and creation operators. Obviously only the state $|1;\alpha\rangle$ with a single electron-hole pair contributes there. Its wave function $\Phi_\alpha(\vec{x}_e;\vec{x}_h) \equiv \langle 0|\psi_e(\vec{x},0\psi_e(\vec{x},0)|1;\alpha\rangle$ satisfies the stationary Schrödinger equation

$$\left\{-\frac{\hbar^2}{2m_e}\nabla_e^2 - \frac{\hbar^2}{2m_h}\nabla_h^2 + E_g - \frac{e^2}{|\vec{x}_e - \vec{x}_h|}\right\}\Phi_\alpha(\vec{x}_e,\vec{x}_h) = E_\alpha\Phi_\alpha(\vec{x}_e,\vec{x}_h).$$

Here we have omitted the hole band index μ since nothing depends on it.

As it is well-known, one may separate the center of mass ($\vec{X} \equiv \frac{m_e\vec{x}_e+m_h\vec{x}_h}{m_e+m}$) motion from the relative one ($\vec{x} \equiv \vec{x}_e - \vec{x}_h$). The former is a free motion defined by a wave vector $\vec{q}$, while the later is characterized by the angular momentum quantum numbers l,m and the energy ε

$$\Phi_{\vec{q},l,m,\varepsilon}(\vec{X},\vec{x}) = \frac{1}{\sqrt{\Omega}}e^{-\imath\vec{q}\vec{X}}\phi_{l,m,\varepsilon}(\vec{x}).$$

The energy ε itself may belong to the discrete hydrogen-like spectrum (bound states) or to the continuum. The total energy is $E_{\vec{q},l,m,\varepsilon} = \frac{\hbar^2q^2}{2(m_e+m_h)} + E_g + \varepsilon$.

Let us insert these ingredients in Eq. (14.1). The time dependence is given by the energy $E_{\vec{q},l,m,\varepsilon}$ of the intermediate state, and since the electron and hole are created in the same point, the wave-function of the relative motion appears only in the origin. To simplify the expressions, we omit mentioning for a while the adiabatic and thermodynamic limits. Then, taking into account also that only hole band-diagonal (identical) terms may contribute, the conductivity tensor is diagonal

$$\sigma(\omega) = -\frac{d^2\omega}{\hbar\Omega}\int_{-\infty}^{0}dt e^{\imath\omega t}\int d\vec{x}\sum_{\vec{q},l,m,\varepsilon}|\phi_{l,m,\varepsilon}(0)|^2\left(e^{\frac{\imath}{\hbar}E_{q,l,\vec{m},\varepsilon}t}e^{\imath\vec{q}\vec{x}} - c.c.\right).$$

Actually in the origin only the $l = 0$ state contributes, and thus after performing the integral over time and momentum (in the infinite volume limit), one gets

$$\sigma(\omega) = \imath \frac{d^2\omega}{\hbar} \sum_\varepsilon |\phi_{0,0,\varepsilon}(0)|^2 \left(\frac{1}{\hbar\omega - E_g - \varepsilon + \imath 0} - c.c. \right).$$

The absorption is given by the real part of $\sigma(\omega)$ for positive frequencies:

$$\Re\sigma(\omega) = \frac{d^2\omega}{\hbar} \sum_\varepsilon |\phi_{0,0,\varepsilon}(0)|^2 \delta(\hbar\omega - E_g - \varepsilon).$$

Introducing the well-known hydrogen-like eigenfunctions and eigenvalues (including those of the continuum) one gets finally the Elliott formula

$$\Re\sigma(\omega) = \frac{d^2\omega}{\pi a_B^3 E_R} \left[\sum_{n=1}^\infty \frac{1}{n^3} \delta(\Delta + \frac{1}{n^2}) + \frac{1}{2}\theta(\Delta) \frac{1}{1 - e^{-\frac{2\pi}{\sqrt{\Delta}}}} \right],$$

where the exciton Rydberg (binding energy of the ground state) E_R and the exciton Bohr radius a_B were introduced, and $\Delta \equiv \frac{\hbar\omega - E_g}{E_R}$. The discrete sum describes the contribution of the bound states (exciton), while the last term is that of the continuous spectrum.

A more sophisticated approach would give rise to a similar expression for the wave-vector dependent conductivity $\sigma(\omega, \vec{k})$ with the only modification $\Delta \to \frac{\hbar\omega - E_g - \frac{\hbar^2 k^2}{2(m_e + m_h)}}{E_R}$. Obviously this takes into account the fact that the absorbed photon momentum imposes also a kinetic energy to the center of mass of the created electron-hole pair.

It is interesting to remark the specific effects of the Coulomb attraction between the created electron and hole. In the limit of no attraction, corresponding to $E_R \to 0$ and $a_B \to \infty$, while $a_B^3 E_R^{\frac{3}{2}} \to \left(\frac{\hbar^2}{2m} \right)^{\frac{3}{2}}$ ($\frac{1}{m} = \frac{1}{m_e} + \frac{1}{m_h}$) one gets the simple inter-band absorption formula

$$\Re\sigma(\omega) = \frac{1}{(2\pi)^2} d^2 \left(\frac{2m}{\hbar^2} \right)^{\frac{3}{2}} \omega\theta(\hbar\omega - E_g)\sqrt{\hbar\omega - E_g}.$$

Besides the absence of the discrete excitonic lines one may remark that while in the Elliot formula the continuum contribution starts with a finite jump, the free electron-hole absorption starts from zero with a square root increase. This effect is known as excitonic enhancement.

In the discussion of the excitonic absorption we did not consider any dissipative effect, therefore the absorption lines are delta-functions. The introduction of the interaction with phonons would introduce a damping and thus a broadening of the excitonic lines.

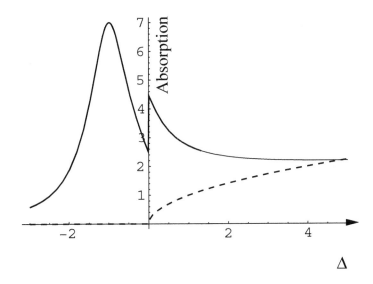

Fig. 14.1 Electron-hole creation by photon absorption with and without Coulomb interaction.

In Fig.(14.1) we illustrate the above mentioned features of the excitonic absorption (here the delta-functions were arbitrarily broadened). The dotted line shows the band to band absorption in the absence of Coulomb effects.

It is important to mention here that our earlier warning - concerning the validity of the Kubo formula in the presence of Coulomb interactions - played no role in this case. Here the longitudinal field, due to internal Coulomb forces is irrelevant. The renormalization of the transverse field is taken into account by the solution of the Maxwell equations.

Chapter 15

The longitudinal dielectric function of crystals

Until now we always assumed that the system under discussion is spatially homogenous. One often constructs through various approximations such models, even for crystalline solids. (Within the approximation of simple parabolic bands in the modeling of a semiconductor, for example.) At a certain macroscopic level surely even a crystal may be regarded as a homogeneous (but anisotropic) medium. Nevertheless, a better understanding of these aspects, as well as a more profound treatment of the linear response of solids, needs a very careful treatment. In this chapter we restrict ourselves to the problem of the macroscopic longitudinal dielectric function $\epsilon(\vec{k}, \omega)$ for Coulomb interacting electrons in a given periodical potential at $T = 0$.

In a homogeneous medium the dielectric tensor may be defined by the relationship between the Fourier-transforms of the total and external potentials

$$\epsilon(\vec{k}, \omega) V^{tot}(\vec{k}, \omega) = V^{ext}(\vec{k}, \omega).$$

In a crystal, however, only a discrete translational invariance holds, and therefore one may define at most a dielectric matrix $\hat{\epsilon}(\vec{k}, \omega)_{\vec{K}\vec{K}'}$ with $\vec{k}$ belonging to the Brillouin zone (BZ) while $\vec{K}$ and $\vec{K}'$ run through the reciprocal lattice

$$\sum_{\vec{K}'} \hat{\epsilon}(\vec{k}, \omega)_{\vec{K}\vec{K}'} V^{tot}(\vec{k} + \vec{K}', \omega) = V^{ext}(\vec{k} + \vec{K}, \omega).$$

Now, in order to define a macroscopic dielectric function one may consider that the external potential is very slow and does not vary in the elementary cell, i.e. $V^{ext}(\vec{k} + \vec{K}, \omega) = 0$ for $\vec{K} \neq 0$. On the other hand, one measures only the macroscopical part of $V^{tot}(\vec{k} + \vec{K}, \omega)$ i.e. only $V^{tot}(\vec{k}, \omega)$.

Then [Adler (1962); Wiser (1963)] it follows the definition

$$\epsilon(\vec{k}, \omega) = \frac{1}{\left[\hat{\epsilon}(\vec{k}, \omega)^{-1}\right]_{00}} . \tag{15.1}$$

The specific aspect of the diagrams of perturbation theory at $T = 0$ in a periodic potential (see [Kirzhnits (1967)]) is that one has to use the basis of the Bloch functions $\phi_{n,\vec{p}}(\vec{x})$, which are the eigenfunctions of the self-consistent (Hartree-Fock) problem in the absence of external fields. That means one has already taken into account - at least at a self-consistent level - that the periodic potential is not just that of the bare ions, i.e. the unperturbed problem is already a self-consistent one. The Green functions will be indexed not only by the conserved wave vector $\vec{p}$, but also by two band indices n, m. The renormalization of the potential line occurs through the summation of the polarization bubbles (self-energies of the longitudinal photon), however, the relationship is either nonlocal in real space or a matrix relation in the Bloch indices. Actually it is more convenient to use the Bloch theorem and get a matrix formulation in the space of the reciprocal lattice. Then the dielectric matrix is related to the polarization part Π by

$$\hat{\epsilon}(\vec{k}, \omega)_{\vec{K}\vec{K}'} = \delta_{\vec{K}\vec{K}'} - \frac{4\pi e^2}{v|\vec{k} + \vec{K}|^2} \sum_{\Gamma, \Gamma'} \langle \vec{K}|u_\Gamma(\vec{k})\rangle \Pi(\omega, \vec{k})_{\Gamma\Gamma'} \langle u_{\Gamma'}(\vec{k})|\vec{K}'\rangle . \tag{15.2}$$

Here v is the volume of the elementary cell and we have introduced the simplifying notation

$$\langle \vec{K}|u_\Gamma(\vec{k})\rangle = \int_v d\vec{x} e^{i(\vec{k}+\vec{K})\vec{x}} \phi_{n\vec{p}}(\vec{x})^* \phi_{m\vec{p}-\vec{k}}(\vec{x}) .$$

Inserting this into Eq. (15.2) one gets for the macroscopical dielectric function

$$\epsilon(\vec{k}, \omega) = 1 - \frac{4\pi e^2}{vk^2} \sum_{\Gamma, \Gamma'} \langle 0|u_\Gamma(\vec{k})\rangle \left[\frac{1}{1 - \Pi\zeta}\Pi\right]_{\Gamma\Gamma'} \langle u_{\Gamma'}(\vec{k})|0\rangle , \tag{15.3}$$

with

$$\zeta_{\Gamma\Gamma'} = \frac{4\pi e^2}{vk^2} \sum_{\vec{K}\neq 0} \frac{\langle u_\Gamma(\vec{k})|\vec{K}\rangle \langle \vec{K}|u_\Gamma(\vec{k})\rangle}{(\vec{k} + \vec{K})^2} .$$

Until now we did not precise the polarization part. Let us consider here the summation of all the ladder diagrams, which is equivalent to the time-dependent Hartree-Fock approximation. Then

$$\Pi(\vec{k},\omega)^{HF}_{\Gamma\Gamma'} = -2\left[\frac{1}{1-\mathcal{K}(\vec{k},\omega)\mathcal{C}^e(\vec{k})}\mathcal{K}(\vec{k},\omega)\right]_{\Gamma\Gamma'},$$

(the factor 2 coming from the spin degeneracy), with the notations

$$\mathcal{K}(\vec{k},\omega)_{\Gamma\Gamma'} = -\delta_{\Gamma\Gamma'}\frac{\theta(\mu-E_{n,\vec{p}})-\theta(\mu-E_{m,\vec{p}-\vec{k}})}{E_{n,\vec{p}}-E_{m,\vec{p}-\vec{k}}+\hbar\omega-\imath 0},$$

where $E_{n,\vec{p}}$ is the energy in the band n and

$$\mathcal{C}^e(\vec{k})_{\Gamma\Gamma'} = \int_v d\vec{x}\int_v d\vec{x}'\phi_{n\vec{p}}(\vec{x})\phi_{m\vec{p}-\vec{k}}(\vec{x}')^*\frac{e^2}{|\vec{x}-\vec{x}'|}\phi_{n'\vec{p}'}(\vec{x})^*\phi_{m'\vec{p}'-\vec{k}}(\vec{x}'),$$

is a sort of exchange Coulomb matrix.

Introducing all this into Eq. (15.3) one finds [Bányai and Gartner (1984)] finally for the macroscopic longitudinal dielectric function of a crystal at $T=0$ within the Hartree-Fock approximation

$$\epsilon(\vec{k},\omega)^{HF}=1-\frac{8\pi e^2}{vk^2}\sum_{\Gamma,\Gamma'}\langle 0|u_\Gamma(\vec{k})\rangle\left[\frac{1}{1+\mathcal{K}(\vec{k},\omega)\left(2\zeta-\mathcal{C}^e(\vec{k})\right)}\mathcal{K}(\vec{k},\omega)\right]_{\Gamma\Gamma'}\langle u_{\Gamma'}(\vec{k})|0\rangle.$$

This is still a very complicated formula, which to our knowledge, has never been used to definite estimations. By neglecting the direct and exchange Coulomb terms from the denominator, one gets a simpler expression, the one often used in phenomenological applications in semiconductor physics, and is directly obtainable from the Kubo formula

$$\epsilon(\vec{k},\omega)^{Kubo}=1+\frac{8\pi e^2}{vk^2}\sum_{n,m,\vec{p}}\frac{\theta(\mu-E_{n,\vec{p}})-\theta(\mu-E_{m,\vec{p}-\vec{k}})}{E_{n,\vec{p}}-E_{m,\vec{p}-\vec{k}}+\hbar\omega-\imath 0}\langle 0|u_{n,\vec{p}}(\vec{k})\rangle\langle u_{m\vec{p}}(\vec{k})|0\rangle.$$

Chapter 16

Charged particles interacting with the quantized electromagnetic field

The theory of condensed matter works mainly with simplified model Hamiltonians; the justification of these models is a separate and seldom well understood task. Nevertheless, it is important to keep in mind that condensed matter is made up of charged particles interacting with the electromagnetic field having a coupled dynamics, and its description must include not only the quantum mechanical description of particles, but also the quantum mechanical nature of the electromagnetic field, i.e. its corpuscular aspects (photons). As the basic picture stems from the classical theory of Lorentz, we shall sketch that first. Thereafter we want to describe shortly the non-relativistic quantum mechanical version and its various, frequently used approximations. This chapter ends with the theory of the transverse dielectric function within this basic frame.

16.1 Classical microscopical theory

The classical microscopical Maxwell-Lorentz equations for the microscopical electrical and magnetic fields $\vec{e}$ and $\vec{b}$ are:

$$
\begin{aligned}
\nabla \times \vec{b} &= \tfrac{4\pi}{c}\vec{j} + \tfrac{1}{c}\tfrac{\partial}{\partial t}\vec{e}\,, \\
\nabla \times \vec{e} &= -\tfrac{1}{c}\tfrac{\partial}{\partial t}\vec{b}\,, \\
\nabla \vec{b} &= 0\,, \\
\nabla \vec{e} &= 4\pi\rho\,.
\end{aligned}
\tag{16.1}
$$

The sources must satisfy the continuity equation:

$$
\nabla \vec{j} + \frac{\partial}{\partial t}\rho = 0\,,
$$

and as charge carriers one had classical point-like particles:

$$\rho(\vec{r}, t) = \sum_i e_i \delta(\vec{r} - \vec{r}_i(t)) + \rho^{ext}(\vec{r}, t),$$

$$\vec{j}(\vec{r}, t) = \sum_i e_i \vec{v}_i(t) \delta(\vec{r} - \vec{r}_i(t)) + \vec{j}^{ext}(\vec{r}, t),$$

governed by Newton's equations of motion:

$$m_i \frac{d}{dt} \vec{v}_i = e_i \left(\vec{e}'(\vec{r}_i, t) + \frac{1}{c} \vec{v}_i \times \vec{b}'(\vec{r}_i, t) \right).$$

Here $\vec{e}'$ and $\vec{b}'$ are the fields without self-action.

The external sources determine the external fields $\vec{E}^{ext}(\vec{r}, t)$, and $\vec{B}^{ext}(\vec{r}, t)$, while $\nabla \vec{j}^{ext} + \frac{\partial \rho^{ext}}{\partial t} = 0$.

The micro-macro transition is understood as a supplementary averaging in space and time, as well as over the initial conditions, and is symbolized by $\langle \ldots \rangle$. Thus the macroscopic fields are given by

$$\vec{E} = <\vec{e}>; \quad \vec{B} = <\vec{b}>,$$

satisfying the Maxwell equations

$$\nabla \times \vec{B} = \frac{4\pi}{c}(<\vec{j}> + \vec{j}^{ext}) + \frac{1}{c} \frac{\partial}{\partial t} \vec{E},$$

$$\nabla \times \vec{E} = -\frac{1}{c} \frac{\partial}{\partial t} \vec{B},$$

$$\nabla \vec{B} = 0,$$

$$\nabla \vec{E} = 4\pi(<\rho> + \rho^{ext}). \tag{16.2}$$

In the phenomenological theory of condensed matter $<\vec{j}(\vec{r}, t)>$ and $<\rho(\vec{r}, t)>$ are considered as given functions of $\vec{E}(\vec{r}, t)$ and $\vec{B}(\vec{r}, t)$. For example

$$<\vec{j}(\vec{x}, t)> = \sigma \vec{E}(\vec{x}, t) + \frac{1}{4\pi}(\epsilon - 1) \frac{\partial \vec{E}(\vec{x}, t)}{\partial t},$$

$$<\rho(\vec{x}, t)> = -\frac{1}{4\pi}(\epsilon - 1) \nabla \vec{E}(\vec{x}, t).$$

We have not introduce the fields $\vec{D}$ and $\vec{H}$ or $\vec{P}$ and $\vec{M}$, since they are not fundamental entities, and have no unambiguous definitions. They are superfluous entities.

16.2   Non-relativistic quantum electrodynamics

A proper treatment of the particle-photon system may be formulated within the frame of (relativistic) quantum-electrodynamics with $\vec{e}, \vec{b}, \vec{j}, \rho$ all operators. However, the most frequently used scheme for condensed matter is to consider the particles within the frame of non-relativistic quantum mechanics, and to quantize the two independent transverse modes of the electromagnetic field as bosons (photons).

In order to formulate this theory, one introduces before quantization, the electromagnetic potentials:

$$\vec{b} = \nabla \times \vec{a},$$

$$\vec{e} = -\nabla v - \frac{1}{c}\frac{\partial}{\partial t}\vec{a}.$$

By these definitions the sourceless Maxwell equations are automatically fulfilled. The only equations that remain are

$$-\nabla^2\vec{a} + \nabla(\nabla\vec{a}) + \frac{1}{c^2}\frac{\partial^2}{\partial t^2}\vec{a} = \frac{4\pi}{c}\vec{j} - \frac{1}{c}\nabla\frac{\partial}{\partial t}v,$$

$$-\nabla^2 v - \frac{1}{c^2}\nabla\frac{\partial}{\partial t}\vec{a} = 4\pi\rho.$$

However, the choice of the potentials is not unique, since gauge transformations

$$v \to v + \frac{1}{c}\frac{\partial}{\partial t}\Lambda,$$

$$\vec{a} \to \vec{a} - \nabla\Lambda,$$

do not change the fields. It is convenient to choose the Coulomb gauge:

$$\nabla\vec{a} = 0,$$

which allows to eliminate the scalar potential v in favor of the charge density ρ through the Poisson equation

$$-\nabla^2 v = 4\pi\rho,$$

giving rise to

$$v(\vec{r}, t) = \int d\vec{r}' \frac{\rho(\vec{r}', t)}{|\vec{r} - \vec{r}'|}.$$

The vector potential also may be eliminated in favor of the current density using the wave equation

$$-\nabla^2 \vec{a} + \frac{1}{c^2}\frac{\partial^2}{\partial t^2}\vec{a} = \frac{4\pi}{c}\vec{j} - \frac{1}{c}\nabla\frac{\partial}{\partial t}v \equiv \frac{4\pi}{c}\vec{j}_\perp \,.$$

From the continuity equation it follows that $\vec{j}_\perp$ is the transverse part of the current density

$$\nabla\vec{j}_\perp = 0 \,,$$

or

$$\vec{j}_\perp(\vec{r},t) \equiv \vec{j}(\vec{r},t) + \frac{1}{4\pi}\nabla\int d\vec{r}'\frac{\nabla'\vec{j}(\vec{r}',t)}{|\vec{r}-\vec{r}'|}\,.$$

The elimination of the vector potential in favor of the transverse current density $\vec{j}_\perp$ and the transverse external e.m. field includes a retardation

$$\vec{a}(\vec{r},t) = \int d\vec{r}'\frac{\vec{j}_\perp(\vec{r}',t-\frac{1}{c}|\vec{r}-\vec{r}'|)}{|\vec{r}-\vec{r}'|}\,,$$

and therefore it is not useful. Instead, one should consider the transverse vector potential $\vec{a}$ as an independent dynamic variable.

Now, we are ready to quantize the theory according to the Hamiltonian prescriptions. Within the second quantization formalism the particle charge density operator is given in terms of the annihilation operator $\psi_{\alpha,\sigma_z}(\vec{r},t)$ of a particle of charge e_α, mass m_α and spin projection $\sigma_z = \pm\frac{1}{2}$ in the point $\vec{r}$ by

$$\rho(\vec{r},t) = \sum_{\alpha,\sigma_z} e_\alpha \psi_{\alpha,\sigma_z}(\vec{r},t)^+\psi_{\alpha,\sigma_z}(\vec{r},t)\,,$$

while the particle current density operator is given by

$$\vec{j}(\vec{r},t) = \sum_{\alpha,\sigma_z}\frac{e_\alpha}{2m_\alpha}\psi_{\alpha,\sigma_z}(\vec{r},t)^+\left(-\imath\hbar\nabla + \frac{e_\alpha}{c}\vec{a}(\vec{r})\right)\psi(\vec{r},t) + h.c.\,,$$

and the (fermionic) commutation relations hold

$$\left[\psi_{\alpha,\sigma_z}(\vec{r},t),\psi_{\alpha',\sigma_z'}(\vec{r}',t)^+\right]_+ = \delta_{\alpha\alpha'}\delta_{\sigma_z\sigma_z'}\delta(\vec{r}-\vec{r}')\,,$$

$$\left[\psi_{\alpha,\sigma_z}(\vec{r},t),\psi_{\alpha',\sigma_z'}(\vec{r}',t)\right]_+ = 0\,.$$

The second quantized wave-functions $\psi_{\alpha,\sigma_z}(\vec{r})$ may be of course decomposed in a superposition of the annihilation operators $a_{\alpha,\vec{q},\sigma}$ of particles with momentum $\vec{q}$ and spin projection σ_z as $\psi_{\alpha,\sigma_z}(\vec{r}) = \frac{1}{\sqrt{\Omega}}\sum_{\vec{k}} e^{-\imath \vec{k}\vec{r}} a_{\alpha,\vec{q},\sigma}$.

The quantization of the transverse electromagnetic field (radiation field), which we denote now by $\vec{A}$, proceeds through the introduction of the two independent transverse photon modes of wave vector $\vec{q}$ (with periodical or with cavity-type boundary conditions)

$$\vec{A}(\vec{x}) = \sum_{\lambda=1,2}\sqrt{\frac{hc}{\Omega}}\sum_{\vec{q}}\frac{1}{\sqrt{|\vec{q}|}}\vec{e}_{\vec{q}}^{(\lambda)}\begin{cases} e^{\imath\vec{q}\vec{x}}\left(b_{\vec{q},\lambda} + b_{-\vec{q},\lambda}^+\right) & \textit{for periodical box}, \\ \imath\sin(\vec{q}\vec{x})\left(b_{\vec{q},\lambda} - b_{\vec{q},\lambda}^+\right) & \textit{for cavity}. \end{cases}$$

Here $\omega_{\vec{q}} = c|\vec{q}|$ is the photon frequency, the two independent (real) polarizations $\vec{e}_{\vec{q}}^{(\lambda)}$ satisfy the relations

$$\vec{q}\vec{e}_{\vec{q}}^{(\lambda)} = 0; \qquad \vec{e}_{\vec{q}}^{(\lambda)}\vec{e}_{\vec{q}}^{(\lambda')} = \delta_{\lambda\lambda'}; \qquad \vec{e}_{\vec{q}}^{(\lambda)} = \vec{e}_{-\vec{q}}^{(\lambda)}; \qquad (\lambda,\lambda' = 1,2),$$

and the photon creation and annihilation operators satisfy the bosonic commutation rules

$$\left[b_{\vec{q},\lambda}, b_{\vec{q}',\lambda'}^+\right] = \delta_{\vec{q},\vec{q}'}\delta_{\lambda\lambda'}; \; [b_{\vec{q},\lambda}, b_{\vec{q}',\lambda'}] = 0.$$

Another useful basis for the photons at a given momentum $\vec{q}$ are the helicity states. The helicity is defined as the sum of the "spins" of the photons along their momenta

$$S = \imath\hbar\int d\vec{q}\left(b_{\vec{q},1}^+ b_{\vec{q},2} - b_{\vec{q},2}^+ b_{\vec{q},1}\right).$$

The operator $b_{\vec{q},1}^+ b_{\vec{q},2} - b_{\vec{q},2}^+ b_{\vec{q},1}$ may be diagonalized simultaneously with $b_{\vec{q},1}^+ b_{\vec{q},1} + b_{\vec{q},2}^+ b_{\vec{q},2}$ (the number of the photons of momentum $\vec{q}$) by the canonical transformation $b_{\vec{q}\pm} = \frac{1}{\sqrt{2}}(b_{\vec{q},1} \pm \imath b_{\vec{q},2})$, giving rise to $b_{\vec{q}+}^+ b_{\vec{q}+} - b_{\vec{q}-}^+ b_{\vec{q}-}$ respectively $b_{\vec{q}+}^+ b_{\vec{q}+} + b_{\vec{q}-}^+ b_{\vec{q}-}$. Therefore, the total helicity is the difference of the numbers of photons of the two helicities, while the total number of photons is the sum of the numbers of photons of the two helicities. The Hamiltonian of the interacting charged fermions and photons is then given by

$$H_{fp} = \sum_{\vec{q},\lambda} \hbar\omega_{\vec{q}} b^+_{\vec{q},\lambda} b_{\vec{q},\lambda} \tag{16.3}$$

$$+ \sum_\alpha \sum_{\sigma=\pm 1} \int d\vec{x} \psi^+_{\alpha,\sigma}(\vec{x}) \left[\frac{1}{2m_\alpha} \left(\frac{\hbar}{\imath}\nabla - \frac{e_\alpha}{c}\vec{A}(\vec{x})\right)^2\right] \psi_{\alpha,\sigma}(\vec{x})$$

$$+ \sum_{\alpha,\alpha'} \sum_{\sigma,\sigma'=\pm 1} \frac{1}{2} \int d\vec{x} \psi^+_{\alpha,\sigma}(\vec{x}) \psi^+_{\alpha,\sigma'}(\vec{x}') \frac{e_\alpha e_{\alpha'}}{|\vec{x}-\vec{x}'|} \psi_{\alpha,\sigma'}(\vec{x}')\psi_{\alpha,\sigma}(\vec{x}) \,,$$

and the equations of motion are the Heisenberg ones.

In the presence of external sources $\rho(\vec{x},t)^{ext}$, $\vec{j}^{ext}(\vec{x},t)$ one has to add also the new interaction terms

$$H_{ext}(t) = -\frac{1}{c} \int d\vec{x} \vec{A}(\vec{x}) \vec{j}^{ext}(\vec{x},t) \tag{16.4}$$

$$+ \sum_{\alpha=e,h} \sum_{\sigma=\pm 1} \int d\vec{x} \int d\vec{x}' \psi^+_{\alpha,\sigma}(\vec{x}) \psi_{\alpha,\sigma}(\vec{x}) \frac{e_\alpha}{|\vec{x}-\vec{x}'|} \rho(\vec{x},t)^{ext} \,.$$

The microscopical Maxwell equation for the vector potential

$$-\nabla^2\vec{A} + \frac{1}{c^2}\frac{\partial^2}{\partial t^2}\vec{A} = \frac{4\pi}{c}\vec{j}_T \,,$$

with the total transverse current $\vec{j}_T$ follows from the Heisenberg equations of motion with the total Hamiltonian

$$H(t) = H_{fp} + H_{ext}(t) \,. \tag{16.5}$$

If we want to take into account also the interaction of the spin with the magnetic field (although its origin is based on the relativistic Dirac equation), we should add a piece

$$\int d\vec{r} \vec{M}(\vec{r}) \vec{B}(\vec{r},t)$$

to the Hamiltonian as well as a piece

$$c\nabla \times \vec{M}(\vec{r})$$

to the current density determined by the spin magnetization density

$$\vec{M}(\vec{r}) \equiv \mu_B \sum_\alpha \sum_{\sigma,\sigma'} \psi_{\alpha,\sigma}(\vec{r})^+ \vec{\sigma}_{\sigma\sigma'} \psi_{\alpha,\sigma'}(\vec{r}) .$$

Here $\mu_B = \frac{e\hbar}{2mc}$ is the Bohr magneton.

Often one ignores completely the radiation field. In this case one has a simple Coulomb interacting particle system, interacting also with a classical external field. A better approximation in some problems, where one cannot neglect the internal magnetic field, is to consider the radiation field at least in a self-consistent (mean-field) approximation, without retardation

$$\vec{A}(\vec{r},t) \approx \vec{A}^{int}(\vec{r},t) = \int d\vec{r'} \frac{\langle \vec{j}_\perp(\vec{r'},t)\rangle}{|\vec{r}-\vec{r'}|} .$$

This means that we consider the interaction of the charged particles only with a s.c. transverse classical field consisting of an internal and an external part $(\vec{A} = \vec{A}^{int} + \vec{A}^{ext})$. Within this s.c. scheme the total energy

$$H + \frac{1}{8\pi} \int d\vec{x} \left(\vec{B}^{int}(\vec{x})^2 + \vec{E}^{int}(\vec{x})^2 \right)$$

is conserved in stationary external fields. (Here $\vec{B}^{int} = \nabla \times \vec{A}^{int}$ and $\vec{E}^{int} = -\frac{1}{c}\frac{\partial}{\partial t}\vec{A}^{int}$.) This improved version has an important deficiency: as any self-consistent theory, it does not fulfill the superposition principle. Nevertheless, the description of most of the electromagnetic experiences with condensed matter is based on the Maxwell equations with quantum-mechanically averaged sources. In the treatment of the optical properties wave propagation is an essential aspect and one uses another self-consistent treatment, in which the field acting on the particles is taken as the implicit solution of the Maxwell equations.

16.3   The transverse dielectric function in the nonrelativistic quantum electrodynamics

Let us consider a system of nonrelativistic charged particles interacting with the quantized electromagnetic field and with external sources, as it was described before. Provided the system (in the absence of the external sources) is homogenous and isotropic, one can define here properly also a transverse dielectric function. Like in the classical Lorentz theory one interprets the classical (macroscopical) fields and potentials ($\vec{E}_{cl} = -\nabla V_{cl} - \frac{1}{c}\frac{\partial}{\partial t}\vec{A}_{cl}$ and $\vec{B}_{cl} = \nabla \times \vec{A}_{cl}$) as averages of the microscopical entities. However, the averaging now means the quantum statistical averaging over the state of the whole system (particles and field). These classical fields are determined by the classical sources

$$\vec{j}_{cl}(\vec{x}) \equiv \vec{j}^{ext}(\vec{x}, t) + \langle\vec{j}(\vec{x})\rangle_t; \quad \rho^{cl}(\vec{x}, t) \equiv \rho^{ext}(\vec{x}, t) + \langle\rho(\vec{x})\rangle_t.$$

Now let us consider a situation where a small deviation from equilibrium occurs, due to the introduction of a weak transversal external field $\vec{A}_{ext}(\vec{x}, t)$ created by a weak external transverse current $\vec{j}^{ext}(\vec{x}, t)$. (The longitudinal case was discussed earlier in Chapter 11 and the present frame does not modify the results.) We shall compare the equation for $\vec{A}_{cl}(\vec{x}, t)$ we derive from the above definitions with the phenomenological, macroscopic Fourier transformed Maxwell equation in a homogenous isotropic medium

$$\left(\epsilon_T(\omega, k)\frac{\omega^2}{c^2} - k^2\right)\tilde{\vec{A}}^{cl}(\omega, k) = -\frac{4\pi}{c}\tilde{\vec{j}}^{ext}_T(\omega, k), \qquad (16.6)$$

in order to define the transverse dielectric function.

Since in equilibrium, in the absence of the external transversal field, the average of the photon creation operator was vanishing - according to linear response theory - to first order in H^{ext} we get

$$\langle A_\mu(\vec{x})\rangle_t = \frac{\imath}{\hbar}\int_{-\infty}^{t} dt'\langle[A_\mu(\vec{x}, t)_H, H^{ext}(t')_H]\rangle$$

$$= \frac{1}{\imath\hbar c}\int_{-\infty}^{t} dt'\int d\vec{x}'\langle[A_\mu(\vec{x}, t)_H, A_\nu(\vec{x}', t')_H]\rangle j^{ext}_\nu(\vec{x}', t')$$

$$= -\frac{1}{c}\int_{-\infty}^{\infty} dt\int d\vec{x}' D^{ret}_{\mu\nu}(\vec{x}, \vec{x}'; t - t')j^{ext}_\nu(\vec{x}', t').$$

Or, in Fourier space, taking into account that $D^{ret}_{\mu\nu}(\omega, \vec{k}) = \left(\delta_{\mu\nu} - \frac{k_\mu k_\nu}{k^2}\right)D^{ret}_T(\omega, \vec{k})$ and the transversality of the external current,

$$\langle A_\mu(\omega, \vec{k})\rangle = -\frac{1}{c}D^{ret}_{\mu\nu}(\omega, \vec{k})j^{ext}_\nu(\vec{k}, \omega) = -\frac{1}{c}D^{ret}_T(\omega, \vec{k})j^{ext}_\mu(\vec{k}, \omega).$$

Therefore, comparing to the classical (macroscopic) Eq. (16.6), one has to identify

$$\epsilon_T(\omega, k)\frac{\omega^2}{c^2} - k^2 = 4\pi D_T^{ret}(\omega, k)^{-1} .$$

If we introduce the retarded photon self-energy (polarization function)

$$D_T^{ret}(\omega, k) = \frac{1}{D_{0T}^{ret}(\omega, k)^{-1} + \Pi_T^{ret}(\omega, k)} ,$$

then with $D_{0T}^{ret}(\omega, k) = \frac{4\pi}{\frac{\omega^2}{c^2} - k^2}$ for the free photon retarded Green function in vacuum we get a simple relation between the transverse dielectric function and the retarded photon polarization part

$$\epsilon_T(\omega, k) = 1 + 4\pi\frac{c^2}{\omega^2}\Pi_T^{ret}(\omega, \vec{k}) .$$

In the next Section we are going to calculate this polarization part in the simple one-loop approximation.

16.4 One-loop (RPA) susceptibility

Let us consider simply a single sort of particles without spin. Then, the current, without the diamagnetic term, is

$$\tilde{j}_\mu(\vec{k}) \equiv \int d\vec{x} e^{i\vec{k}\vec{x}} j_\mu(\vec{x}) = \frac{e\hbar}{m}\sum_{\vec{p}} p_\mu a_{\vec{p}+\frac{1}{2}\vec{k}}^+ a_{\vec{p}-\frac{1}{2}\vec{k}} .$$

Within the Keldysh formalism [Keldysh (1965)] (see Chapter 23) one has for the polarization bubble:

$$\Pi^{ret}(t, t'; k) = \theta(t - t') \left[\Pi^>(t, t'; k) - \Pi^<(t, t'; k)\right] .$$

Considering the $\frac{1}{c}\vec{j}\vec{A}$ interaction (without vertex corrections), its transverse part is

$$\Pi_T^{\gtrless}(t, t'; k) = e^2 \frac{i\hbar^2}{2c^2m^2}\frac{1}{\Omega}\sum_{\vec{p}} G_{\vec{p}+\frac{1}{2}\vec{k}}^{\gtrless}(t, t')G_{\vec{p}-\frac{1}{2}\vec{k}}^{\lessgtr}(t', t)\left(p^2 - \frac{(\vec{p}\vec{k})^2}{\vec{k}^2}\right) .$$

The electron Green functions are

$$G_{\vec{p}}^{>}(t,t') = -\imath < a_{\vec{p}}(t)a_{\vec{p}}^{+}(t') > ,$$

$$G_{\vec{p}}^{<}(t,t') = \imath < a_{\vec{p}}^{+}(t')a_{\vec{p}}(t) > .$$

To lowest order in the electromagnetic interaction we consider only the free propagators in equilibrium

$$G_{0\vec{p}}^{>}(t,t') = \imath(f_{\vec{p}} - 1)e^{-\frac{\imath}{\hbar}e_{\vec{p}}(t-t')} ,$$

$$G_{0\vec{p}}^{<}(t,t') = \imath f_{\vec{p}}e^{-\frac{\imath}{\hbar}e_{\vec{p}}(t-t')} ,$$

and with

$$\Pi_{0}^{ret}(\omega,k) \equiv \int_{0}^{\infty} dt e^{-\imath(\omega+\imath 0)t}\Pi_{0}^{ret}(t) ,$$

one gets for the transverse susceptibility of the electron gas

$$\chi(\omega,k)_{T} = \frac{c^2}{\omega^2}\Pi^{ret}(\omega,\vec{k}) \tag{16.7}$$

$$= e^2 \frac{\hbar^2}{2\omega^2 m^2} \frac{1}{\Omega} \sum_{\vec{p}} \frac{f_{\vec{p}-\frac{1}{2}\vec{k}} - f_{\vec{p}+\frac{1}{2}\vec{k}}}{\hbar\omega - e_{\vec{p}+\frac{1}{2}\vec{k}} + e_{\vec{p}-\frac{1}{2}\vec{k}} + \imath 0}\left(p^2 - \frac{(\vec{pk})^2}{\vec{k}^2}\right) .$$

The transverse susceptibility Eq.(16.7) has to be compared with the corresponding expression for the RPA longitudinal susceptibility (Lindhard formula)

$$\chi(\omega,k)_{L} = \frac{e^2}{k^2} \frac{1}{\Omega} \sum_{\vec{p}} \frac{f_{\vec{p}+\frac{1}{2}\vec{k}} - f_{\vec{p}-\frac{1}{2}\vec{k}}}{\hbar\omega - e_{\vec{p}+\frac{1}{2}\vec{k}} + e_{\vec{p}-\frac{1}{2}\vec{k}} + \imath 0} .$$

Now, decomposing in Eq.(16.7) the integration momentum $\vec{p}$ into its components transverse and longitudinal $\vec{p} = \vec{p}_T + \vec{p}_L$ to $\vec{k}$ one may bring Eq.(16.7) into the form

$$\chi(\omega,k)_{T} = \frac{e^2\hbar^2}{2m^2\omega^2} \frac{1}{(2\pi)^3} \int d\vec{p}_T \int_{-\infty}^{\infty} dp_L p_T^2 \tag{16.8}$$

$$\times \frac{f\left(p_T^2 + p_L^2 - p_L k + \frac{1}{4}k^2\right) - f\left(p_T^2 + p_L^2 + p_L k + \frac{1}{4}k^2\right)}{\hbar\omega - \frac{\hbar^2}{m}p_L k + \imath 0} .$$

The solutions of the equation

$$\epsilon_T(\omega, k) \frac{\omega^2}{c^2} - k^2 = 0$$

define the transverse plasma eigenmodes, i.e.

$$\epsilon_T(\omega, k) = \frac{k^2 c^2}{\omega^2} ,$$

to be compared with that for longitudinal plasma modes

$$\epsilon_l(\omega, k) = 0 .$$

While in the longitudinal case one gets also for $k = 0$ a solution (with the plasma frequency), in the transverse case there is no solution at all for $k = 0$ since the transverse susceptibility itself vanishes at $k = 0$.

Both susceptibilities, although without dissipation have imaginary parts. In the case of the longitudinal susceptibility, this leads to the well-known Landau damping of the plasmon modes. In the case of the transverse susceptibility however one has to consider wave vectors of order $\frac{\omega}{c}$. At these wave vectors, as it may be seen from Eq.(16.8) the imaginary part comes from the contribution to the integral of longitudinal electron velocities $\frac{\hbar}{m} p_L$ of the order of the light velocity c. Since our model for electrons is non-relativistic, this domain is not physical. On the other hand, relativistic energy and momentum conservation with free particles forbids also such an absorption process in which a photon is absorbed by a free electron. It would require $\hbar\omega + \hbar\sqrt{\left(\frac{mc^2}{\hbar}\right)^2 + c^2\vec{p}^2} = \hbar\sqrt{\left(\frac{mc^2}{\hbar}\right)^2 + c^2(\vec{p}+\vec{k})^2}$. In the rest frame of the electron $\vec{p} = 0$ and taking into account the relationship $k = \frac{\omega}{c}$, this leads to the equation $\omega + \frac{mc^2}{\hbar} = \sqrt{\omega^2 + \left(\frac{mc^2}{\hbar}\right)^2}$, which has only the solution $\omega = 0$ Nevertheless, the process is allowed if either energy non-conservation (ultrashort-processes) or momentum non-conservation (in the presence of an external potential (like in a crystal) occurs. Since the light velocity in medium is also much beyond the velocities of the nonrelativistic electrons in the medium, the argument is valid also when c is the light velocity in the medium.

Chapter 17

Master equations for the density operator (Reservoir theory)

In this Chapter we want to describe the derivation of the Master equation from the quantum mechanics of the system proper A interacting with a reservoir (thermostat) B. Our more ambitious purpose is to derive not only the already discussed Master equation for the diagonal matrix elements of the density operator (probabilities) for the system A, but also that of describing the evolution of the off-diagonal ones. This would enable the calculation of the average of an arbitrary operator, therefore essentially enlarging the domain of applications.

The Hamiltonian of the whole system is the sum of the Hamiltonians of the systems A and B and of their interaction:

$$H = H_A + H_B + H_{AB} .$$

The eigenstates and eigenenergies of the two independent systems are supposed to be known

$$H_A|\alpha_A\rangle = \epsilon_{\alpha_A}|\alpha_A\rangle; \quad H_B|\alpha_B\rangle = \epsilon_{\alpha_B}|\alpha_B\rangle .$$

The quantum-statistical average of an operator ϑ_A acting in the space of eigenstates of H_A is given by

$$\langle\vartheta_A\rangle = Tr\,(\rho\vartheta_A) = \sum_{\alpha_A,\alpha_B} \langle\alpha_A|\langle\alpha_B|\rho\vartheta_A|\alpha_B\rangle|\alpha_A\rangle$$

$$= \sum_{\alpha_A}\langle\alpha_A| \left(\sum_{\alpha_B}\langle\alpha_B|\rho|\alpha_B\rangle\right) \vartheta_A|\alpha_A\rangle$$

$$= Tr_A\,(\rho_A\vartheta_A) ,$$

where the reduced density operator of the system A

$$\rho_A \equiv Tr_B \rho$$

was introduced. The focus of our interest is this operator. Let us consider an initial state at $t = 0$ with factorizable density operator (independent systems with no initial correlations)

$$\rho(0) = \rho_A(0)\rho_B(0) . \tag{17.1}$$

The reservoir B is supposed to have been in thermal equilibrium described by the macro-canonical distribution

$$\rho_B(0) = \rho_B^0(H_B) \equiv \frac{e^{-\beta(H_B - \mu_B N_B)}}{Z} ,$$

while the system A is in an arbitrary initial state $\rho_A(0)$.

From this moment on ($t > 0$) the Liouville equation

$$i\hbar \frac{\partial}{\partial t}\rho(t) = [H, \rho(t)]$$

describes the evolution of the whole system A+B.

In the interaction picture

$$\tilde{\rho}(t) \equiv e^{\frac{i}{\hbar}(H_A + H_B)t}\rho(t)e^{-\frac{i}{\hbar}(H_A + H_B)t} ,$$

the Liouville equation reads as

$$i\hbar \frac{\partial}{\partial t}\tilde{\rho}(t) = \left[\tilde{H}_{AB}(t), \tilde{\rho}(t)\right] ,$$

with

$$\tilde{H}_{AB}(t) \equiv e^{\frac{i}{\hbar}(H_A + H_B)t} H_{AB} e^{-\frac{i}{\hbar}(H_A + H_B)t} .$$

The integral form of this equation is

$$\tilde{\rho}(t) = \tilde{\rho}(0) + \frac{1}{i\hbar} \int_0^t dt' \left[\tilde{H}_{AB}(t'), \tilde{\rho}(t')\right] .$$

We will use it in the right hand side of the original Liouville equation to get

$$\frac{\partial}{\partial t}\tilde{\rho}(t) = \frac{1}{\imath\hbar}\left[\tilde{H}_{AB}(t),\tilde{\rho}(0)\right] + \frac{1}{(\imath\hbar)^2}\int_0^t dt'\left[\tilde{H}_{AB}(t),\left[\tilde{H}_{AB}(t'),\tilde{\rho}(t')\right]\right].$$

Performing a trace over the degrees of freedom of the reservoir B we get for the evolution of the reduced density operator for the system A the equation

$$\frac{\partial}{\partial t}\tilde{\rho}_A(t) = \frac{1}{\imath\hbar}Tr\left\{\left[\tilde{H}_{AB}(t),\tilde{\rho}(0)\right]\right\} \tag{17.2}$$

$$+ \frac{1}{(\imath\hbar)^2}\int_0^t dt'Tr\left\{\left[\tilde{H}_{AB}(t),\left[\tilde{H}_{AB}(t'),\tilde{\rho}(t')\right]\right]\right\},$$

which, however, is not closed for $\tilde{\rho}_A$.

Now we are ready to introduce a series of assumptions and approximations; without it we couldn't move forwards.

The first assumption

$$Tr_B\left(\rho_B^0 H_{AB}\right) = 0 \tag{17.3}$$

eliminates the inhomogeneous term containing the information about the initial state. This condition is usually satisfied, for example by the electron-phonon interaction, due to its linearity in the phonon creation and annihilation operators. Then the equation becomes homogenous

$$\frac{\partial}{\partial t}\tilde{\rho}(t) = \frac{1}{(\imath\hbar)^2}\int_0^t dt'\left[\tilde{H}_{AB}(t),\left[\tilde{H}_{AB}(t'),\tilde{\rho}(t')\right]\right]. \tag{17.4}$$

We want to be even more specific about the interaction Hamiltonian and assume that it has the structure

$$H_{AB} = g\sum_\lambda V_A^{(\lambda)}V_B^{(\lambda)}; \quad V_{A,B}^{(\lambda)+} = V_{A,B}^{(\lambda)}, \tag{17.5}$$

with the hermitian operators $V_{A,B}^\lambda$ acting in the spaces A respectively B. This again is a widely valid assumption.

We admitted the absence of initial correlations, but until now no approximations were made, however, we shall make here the basic *factorization. approximation* We assume that, under the integral, one may replace the density operator for the whole system by the product of the density matrix of the system A and the unperturbed density operator for the reservoir B.

$$\tilde{\rho}(t) \approx \tilde{\rho}_A(t)\rho_B^0 \,. \tag{17.6}$$

This assumption corresponds to the feature a thermostat is supposed to have.

Introducing the form of the interaction given by Eq.(17.5) and the approximation contained in Eq.(17.6) into the right hand side of Eq.(17.4) we get

$$\frac{\partial}{\partial t}\tilde{\rho}_A(t) = -\frac{g^2}{\hbar^2}\sum_{\lambda,\lambda'}\int_0^t d\tau \left\{ G_{\lambda\lambda'}(\tau) \left[\tilde{V}_A^{(\lambda)}(t)\tilde{V}_A^{(\lambda')}(t-\tau)\tilde{\rho}_A(t-\tau) \right. \right.$$
$$\left. \left. - \tilde{V}_A^{(\lambda')}(t-\tau)\tilde{\rho}_A(t-\tau)\tilde{V}_A^{(\lambda)}(t) \right] + h.c. \right\}, \tag{17.7}$$

where

$$G_{\lambda\lambda'}(\tau) \equiv Tr\left\{ \rho_B^0 \tilde{V}_B^{(\lambda)}(\tau)\tilde{V}_B^{(\lambda')}(0) \right\}, \tag{17.8}$$

with the properties

$$G_{\lambda\lambda'}(\tau) = G_{\lambda'\lambda}(-\tau)^* = G_{\lambda\lambda'}(-\tau + \imath\hbar\beta). \tag{17.9}$$

This is still an equation with memory, like in quantum kinetics. We want to get rid of the memory in a reasonable way. In order to do this, we assume that $G_{\lambda\lambda'}(\tau)$ is peaked at $\tau = 0$. This assumption is called usually as that of a "singular reservoir" . Then, under the integral one may replace

$$\tilde{\rho}_A(t - \tau) \approx \tilde{\rho}_A(t) \,, \tag{17.10}$$

and we get a local equation.

Finally, assuming that we are in the asymptotic regime, we extend the integration limit to infinity, i.e.

$$\int_0^t d\tau \to \int_0^\infty d\tau \,.$$

After all these steps we return to the Schrödinger picture getting

$$\frac{\partial}{\partial t}\rho_A(t) = \frac{1}{\imath\hbar}\left[H_A,\rho_A(t)\right] - \frac{g^2}{\hbar^2}\sum_{\lambda,\lambda'}\int_0^\infty d\tau\left\{G_{\lambda\lambda'}(\tau)\left[V_A^{(\lambda)}\tilde{V}_A^{(\lambda')}(-\tau)\rho_A(t)\right.\right.$$

$$\left.\left. - \tilde{V}_A^{(\lambda')}(-\tau)\rho_A(t)V_A^{(\lambda)}\right] + h.c.\right\}. \tag{17.11}$$

This equation can be written in a simplified symbolic form by the introduction of the concept of a "super-operator". A super-operator is an operator, whose matrix elements are linear functionals of the matrix elements of another operator. Accompanied by a change of notations for the matrix elements of the reduced density operator

$$\langle\alpha_A|\rho_A|\alpha_A'\rangle \Longrightarrow \rho_{ij}^A\,; \qquad \epsilon_A \Longrightarrow \epsilon_i\,,$$

we may write

$$\frac{\partial}{\partial t}\rho_{ij}^A(t) = -\frac{\imath}{\hbar}(\epsilon_i - \epsilon_j)\rho_{ij}^A(t) - \sum_{l,m}K_{ij;lm}\rho_{lm}^A(t)\,,$$

or

$$\frac{\partial}{\partial t}\rho^A = -\frac{\imath}{\hbar}\left[H_A,\rho^A\right] - K\rho^A\,,$$

where K is the super-operator defined by $K_{ij;lm}$.

The last approximation we shall introduce is the so called "secular approximation", which consists in omitting terms with rapid free oscillations. This implies keeping only terms with $\epsilon_i - \epsilon_j = \epsilon_l - \epsilon_m$. By implementing this approximation, however, the lack of any degeneracy is supposed, and this is a very restrictive requirement.

To eliminate the non-desired terms, one introduces the projectors on the state $|i\rangle$: $P_i^2 = P_i$. Then we get finally the Master equation for the reduced density operator of the system A

$$\frac{\partial}{\partial t}\rho_A(t) = \frac{1}{\imath\hbar}\left[H_A,\rho_A(t)\right] - K^\natural\rho_A(t)\,, \tag{17.12}$$

with the modified super-operator

$$K^{\natural}\rho_A \equiv \frac{g^2}{\hbar^2} \sum_{\epsilon_i - \epsilon_j = \epsilon_l - \epsilon_m} \sum_{\lambda,\lambda'} \int_0^\infty d\tau \Big\{ G_{\lambda\lambda'}(\tau) P_i \left[V_A^{(\lambda)} \tilde{V}_A^{(\lambda')}(-\tau) P_l \rho_A(t) P_m \right.$$

$$\left. - \tilde{V}_A^{(\lambda')}(-\tau) P_l \rho_A(t) P_m V_A^{(\lambda)} \right] P_j + h.c. \Big\} . \tag{17.13}$$

After all these lengthy and not well controlled approximations, we got an equation which has remarkable properties, deriving from the properties of the super-operator $K^{\natural}$:

  i) $Tr\left(K^{\natural}\rho_A\right) = 0$, thus, the trace of ρ_A is conserved;

  ii) $\left(K^{\natural}\rho_A\right)^+ = \left(K^{\natural}\rho_A\right)$, thus, the hermiticity of ρ_A is conserved;

  iii) $K^{\natural}e^{-\beta(H_A - \mu_A N_A)} = 0$, thus, the macro-canonical distribution is a stationary solution.

Furthermore, it can be shown that the equation conserves the positivity of the density operator

  iv) $\rho_A(0) > 0 \Rightarrow \rho_A(t) > 0$, and its solution tends asymptotically to the unique equilibrium solution

  v) $\rho_A(t) \rightarrow \frac{1}{Z_A} e^{-\beta(H_A - \mu_A N_A)}$ for $t \rightarrow \infty$.

The equation for the diagonal matrix elements coincides with the known diagonal Master equation of Chapter 8 for the probabilities $\mathcal{P}_i(t) \equiv \langle i|\rho_A(t)|i\rangle$:

$$\frac{d}{dt}\rho_{ii}^A(t) = -\sum_j \left(\rho_{ii}^A(t)W_{ij} - \rho_{jj}^A(t)W_{ji}\right) , \tag{17.14}$$

with the transition rates

$$W_{ij} \equiv \frac{2\pi}{\hbar} \sum_{\alpha_B,\alpha'_B} \frac{1}{Z_B} e^{-\beta(H_B - \mu_B N_B)} |\langle\alpha_B|\langle i|H_{AB}|j\rangle|\alpha'_B\rangle|^2 \delta(\epsilon_i + E_{\alpha_B}^B - \epsilon_{j0} + E_{\alpha'_B}^B),$$

given by the "golden rule". Of course, here the thermodynamic limit for the reservoir (a continuous spectrum) is understood, otherwise the above expression, containing a delta-function, would become meaningless.

In the case of the off-diagonal matrix elements ($i \neq j$) there are two kind of terms: with imaginary or with real coefficients. The first ones may be interpreted as second order energy corrections, and we shall absorb them in the definition of the energies. As a result we get

$$\frac{\partial}{\partial t}\rho_{ij}^{A}(t) = -\frac{\imath}{\hbar}(\epsilon_i - \epsilon_j)\rho_{ij}^{A}(t) - \frac{1}{2}\sum_k(W_{ik} + W_{jk})\rho_{ij}^{A}(t); \quad (i \neq j) \quad (17.15)$$

which shows that the off-diagonal matrix elements decay exponentially as it is to be expected by the property v). This last result is intimately related to the absence of initial correlations we admitted.

Chapter 18

Two-level atom interacting with light

The earliest and also simplest application of the Master equation (including the non-diagonal matrix elements) was in the field of quantum optics of atoms. Let us consider the Hamiltonian of a two-level system (atom) in the second quantization formalism:

$$H_0 = e_1 a_1^+ a_1 + e_2 a_2^+ a_2 \,.$$

The need for non-diagonal matrix elements of the density operator comes from the introduction of a time dependent perturbation by a classical electromagnetic pulse $E(t) = \mathcal{E}(t) \cos(\omega t - \phi)$, having the carrier frequency ω and the envelope $\mathcal{E}(t)$. This coherent interaction with the field

$$H'(t) = -dE(t) \left(a_2^+ a_1 + a_1^+ a_2 \right)$$

causes transitions between the two atomic levels. Here we admitted implicitly a dipole interaction with the field and therefore one of the levels must be of s-type, the other one p-type, the quantization axis z being oriented along the field.

One is looking for the density operator of the 2-level system

$$\rho_{ij} = \langle a_i^+ a_j \rangle \ (i, j = 1, 2) \,,$$

with positive diagonal elements

$$\rho_{11}, \rho_{22} \geq 0 \,.$$

Considering only a single electron (on the average), the normalization is

$$\rho_{11} + \rho_{22} = 1.$$

Form the above two pieces of the Hamiltonian the Liouville equations

$$i\hbar \frac{\partial}{\partial t}\rho_{11} = dE(t)(\rho_{21} - \rho_{12}),$$

$$i\hbar \frac{\partial}{\partial t}\rho_{12} = (e_1 - e_2)\rho_{12} + dE(t)(\rho_{22} - \rho_{11}),$$

emerge. Now, we need to introduce the interaction with a thermostat, which, in this case, is supposed to be given by the radiation field, i.e. by the photons in the system. At this step we are oriented by the results of the previous chapter, although they were not derived for the case of time dependent Hamiltonians.

If we include the irreversible terms due to the reservoir (thermostat) as in Eqs.(17.14),(17.15), we get

$$\frac{\partial}{\partial t}\rho_{11} = \frac{1}{\hbar}dE(t)\Im(\rho_{21}) - W_{12}\rho_{11} + W_{21}(1 - \rho_{11}), \qquad (18.1)$$

$$\frac{\partial}{\partial t}\rho_{12} = \frac{1}{i\hbar}(e_1 - e_2)\rho_{12} + \frac{1}{i\hbar}dE(t)(1 - 2\rho_{11}) - \frac{1}{2}(W_{12} + W_{21})\rho_{12},$$

with

$$W_{12} = W_{21}e^{\beta(e_1 - e_2)}.$$

The traditional choice of variables is $\nu \equiv \rho_{11} - \rho_{22}$ the population inversion and $p \equiv \rho_{21}$ the polarization.

At very low temperatures W_{12} ($e_1 < e_2$) may be discarded, and by the introduction of the relaxation time T

$$\frac{1}{T} \equiv W_{21},$$

we get

$$\frac{\partial}{\partial t}\nu(t) = \frac{2}{\hbar}dE(t)\Im\{p(t\} - \frac{1}{T}(\nu(t) + 1),$$

$$\left(\frac{\partial}{\partial t} + \frac{i}{\hbar}(e_2 - e_1)\right)p(t) = -\frac{i}{\hbar}dE(t)\nu(t) - \frac{1}{2T}p(t).$$

Within quantum optics one makes a further useful approximation, namely, neglects the rapidly oscillating part of the inter-band polarization ("rotating wave approximation"). For the population inversion ν and the slowly varying part of the inter-level polarization

$$p(t) \equiv \bar{p}(t)e^{-\imath\omega t}$$

one gets then the Bloch equations

$$\frac{\partial}{\partial t}\nu(t) = 2\Im\left(\Omega(t)^*\bar{p}(t)\right) - \frac{1}{T}(\nu(t)+1)\,, \qquad (18.2)$$

$$\left(\frac{\partial}{\partial t} - \imath\Delta + \frac{1}{2T}\right)\bar{p}(t) = -\frac{\imath}{2}\Omega(t)\nu(t) - \frac{1}{2T}\bar{p}\,,$$

with the detuning

$$\Delta \equiv \frac{1}{\hbar}(e_1 - e_2 + \hbar\omega)\,,$$

and the Rabi frequency

$$\Omega(t) \equiv \frac{d}{\hbar}\mathcal{E}(t)\,.$$

As we can see, the inter-level polarization decay-time is twice more rapid as that of the population inversion. This is partly due to the fact that while the inter-level polarization is a probability amplitude, the inversion is a probability.

For an atomic gas, one should consider for each one of the atoms with the coordinate $\vec{x}$ such an equation with the field $\vec{E}(\vec{x},t)$. One should take into account also the Doppler shift, due to the velocity of the atom $\vec{v}$. Actually we have to write down simultaneously also the Maxwell equations for the e.m. field, whose sources are some external dipoles and the inter-level polarizations of the atoms (induced dipole moments) $\vec{d}(\vec{x})(p(\vec{x},t)+p(\vec{x},t)^*)$ themselves.

With these ingredients the Bloch equations have served successfully for the interpretation of many experiments with atoms irradiated by short laser pulses. These equations were extended further also to the study of ultra rapid optical phenomena in semiconductors, as we shall see it in later chapters.

Chapter 19

Master equations for the density operator with particle-particle collisions (The super-projector approach)

Until now we have discussed only the case where irreversibility comes in through the interaction with a thermostat. The concept of the open system, however, may be extended also to the case without thermostat. In a many-body system, one concentrates only on a few degrees of freedom, trying to obtain closed equations only for these, eliminating higher correlations. This was the case with the classical Boltzmann equation, describing collisions between the particles, where the focus of interest is on the probability to find a particle with a given velocity and coordinate, ignoring higher, correlated probabilities.

Elastic collisions cannot give rise to a canonical distribution and therefore to the concept of a temperature for the whole system, since energy (and momentum) is conserved; nevertheless, for the one-particle probability one may introduce a temperature determined by the average one-particle energy.

As a rule, in the quantum mechanical treatment higher correlations are eliminated, which then represent a thermostat for the one-particle degrees of freedom.

Here we want to describe a more ambitious projection super-operator formalism [Nakajima (1958); Zwanzig (1960)] in order, to obtain closed, Markovian equations for the diagonal and off-diagonal matrix-elements of the density operator (see also [Zwanzig (2001)]). This approach aims again at the description of the particle-particle collisions, which conserve the energy of the system.

This theory has a weak point: namely that one might expect their validity only in the thermodynamic limit, while this very peculiarity in the derivation is hard to take into account explicitly. On the other hand, the seeming generality of the formal result raises more questions about the validity of the obtained results.

Let us consider a Hamiltonian split into its unperturbed (free) part and a perturbation describing particle-particle interaction

$$H = H_0 + H_1 \,.$$

The eigenstates of the unperturbed part

$$H_0 |\nu> = E_\nu |\nu>$$

are supposed to be known. With the help of the projection operator

$$P|\nu><\nu'| = \delta_{\nu\nu'} |\nu><\nu| \,; \qquad (P^2 = P) \,,$$

one may split the density operator into two pieces

$$\rho = \rho_d + \rho_n \,,$$

the diagonal and the off-diagonal parts:

$$\rho_d \equiv P\rho; \qquad \rho_n \equiv (1 - P)\rho \,.$$

The density operator may be written in the basis of H_0 as

$$\rho = \sum_{\nu,\nu'} \rho_{\nu\nu'} |\nu><\nu'| \,,$$

and the action of the super-operator P gives rise to

$$P\rho == \sum_\nu \rho_{\nu\nu} |\nu><\nu| \,.$$

Obviously $1 - P$ produces then the remaining off-diagonal part.

By the introduction of the Liouville super-operator, the Liouville equation for the density operator

$$\imath\hbar\dot\rho = [H, \rho] \,,$$

looks as

$$\imath\hbar\dot\rho = L\rho \,.$$

We may try to split this in the same manner

$$
\begin{cases}
i\hbar P\dot{\rho} = PLP\rho + PL(1-P)\rho\,, \\
i\hbar(1-P)\dot{\rho} = LP\rho + (1-P)L(1-P)\rho\,.
\end{cases}
\tag{19.1}
$$

Now, we try to obtain a closed equation for the diagonal part. The formal solution of the off-diagonal equation is

$$
(1-P)\rho(t) = e^{-\frac{i}{\hbar}(1-P)Lt}(1-P)\rho(0) - \frac{i}{\hbar}\int_0^t dt'\, e^{-\frac{i}{\hbar}(1-P)Lt'}(1-P)LP\rho(t-t')\,.
$$

Inserting it into the equation for the diagonal part $\rho_d \equiv P\rho$ we get

$$
\dot{\rho}_d(t) = -\frac{i}{\hbar}L_{eff}^d\rho_d(t) + \int_0^t dt'\, K^d(t')\rho_d(t-t') + I^d(t)\,,
\tag{19.2}
$$

and with the splitting of the Liouville super-operator into its unperturbed and perturbed parts ($L = L_0 + L_1$) we find

$$
L_{eff}^d = PL = PL_1\,,
$$

$$
K^d(t) = -\frac{1}{\hbar^2}PLe^{-\frac{i}{\hbar}(1-P)Lt}(1-P)LP = -\frac{1}{\hbar^2}PL_1 e^{-\frac{i}{\hbar}(1-P)Lt}(1-P)L_1 P\,,
$$

and

$$
I^d(t) = -\frac{i}{\hbar}PLe^{-\frac{i}{\hbar}(1-P)Lt}(1-P)\rho(0) = -\frac{i}{\hbar}PL_1 e^{-\frac{i}{\hbar}(1-P)Lt}(1-P)\rho(0)\,.
$$

Here we used the property

$$
PL_0 = L_0 P = 0\,,
$$

resulting from the definition of the projector on the eigenstates of the unperturbed Hamiltonian.

As in the case of the reservoir theory, one implements a Markovian approximation by assuming slowly varying asymptotic regime, such that the replacement

$$
\rho_d(t-t') \to \rho_d(t)\,,
$$

under the integral and the the extension of the integration limit

$$\int_0^t dt' \cdots \to \int_0^\infty dt' \cdots$$

may be performed.

Then, to the lowest order in L_1, one has:

$$\dot{\rho}_d(t) = -\frac{\imath}{\hbar} P L_1 \rho_d(t) + \int_0^t dt' K_0^d(t')\rho(t-t') - \frac{\imath}{\hbar} P L_1 e^{-\frac{\imath}{\hbar}L_0 t}\rho_n(0), \qquad (19.3)$$

with

$$K_0^d(t) = -P L_1 e^{-\imath L_0 t}(1-P)L_1 .$$

With these approximations we get

$$\int_0^\infty dt' K_0^d(t')\rho_d(t) = -\frac{1}{\hbar^2} \int_0^\infty dt' \left[H_1, e^{-\frac{\imath}{\hbar}H_0 t'} [H_1, \rho_d]_n e^{\frac{\imath}{\hbar}H_0 t'} \right]_d$$

$$= -\frac{1}{\hbar^2} \int_0^\infty dt' \left[H_1, e^{-\frac{\imath}{\hbar}H_0 t'} [H_1, \rho_d] e^{\frac{\imath}{\hbar}H_0 t'} \right]_d$$

$$= -\frac{1}{\hbar^2} \int_0^\infty dt' \left\{ H_1 e^{-\frac{\imath}{\hbar}H_0 t'} (H_1\rho_d - \rho_d H_1) e^{\frac{\imath}{\hbar}H_0 t'} \right.$$

$$\left. - e^{-\frac{\imath}{\hbar}H_0 t'} (H_1\rho_d - \rho_d H_1) e^{\frac{\imath}{\hbar}H_0 t'} \right\}_d .$$

where the lower index d indicates the diagonal part. Thus,

$$< \nu| \int_0^\infty dt' K_0^d(t')\rho_d(t)|\nu >$$

$$= -\frac{2\pi}{\hbar} \sum_{\nu'} \delta(E_\nu - E_{\nu'})| < \nu|H_1|\nu' > |^2 (< \nu|\rho|\nu > - < \nu'|\rho|\nu' >) ,$$

and we get the collision term for the diagonal part of the density operator

$$\frac{d}{dt} < \nu|\rho|\nu > |_{coll} = -\sum_{\nu'} W_{\nu\nu'} (< \nu|\rho|\nu > - < \nu'|\rho|\nu' >) , \qquad (19.4)$$

with the "golden rule" transition rate

$$W_{\nu\nu'} \equiv \frac{2\pi}{\hbar} \delta(E_\nu - E_{\nu'})| < \nu|H_1|\nu' > |^2 .$$

If the off-diagonal part is vanishing at $t = 0$ ($\rho_n(0) = 0$), and the interaction Hamiltonian has no diagonal part ($< \nu|H_1|\nu >= 0$), then, from Eq. (19.3) we get the known diagonal Master equation

$$\dot{\rho}_d(t) = -\sum_{\nu'} (W_{\nu\nu'} < \nu|\rho|\nu > -W_{\nu'\nu} < \nu'|\rho|\nu' >) . \qquad (19.5)$$

When we wrote the collision term in this standard form, we also took into account that $W_{\nu\nu'} = W_{\nu'\nu}$. It is obvious that the energy conserving delta functions appearing in the transition rates are meaningless in a discrete spectrum. One may expect that the formal result might hold at most in a continuous spectrum. It is to be expected that further limitations must be taken into account. Formally it looks like one could apply this theory also to a single particle in a potential - yet this proves to be a false expectation. The continuous spectrum must be achieved through a thermodynamic limit, in which the volume of the system increases together with the number of particles. The derivation does not touch all these delicate aspects and therefore it is not a real proof.

Now we try to derive a closed equation also for the off-diagonal part of the density operator.

We repeat the same steps with the off-diagonal part ρ_n. We introduce the formal solution of the diagonal part into the equation of the off-diagonal part and get

$$\dot{\rho}_n(t) = -\frac{\imath}{\hbar} L^n_{eff}\rho_n(t) + \int_0^t dt' K^n(t')\rho_n(t - t') + I^n(t), \qquad (19.6)$$

with

$$L^n_{eff} = (1 - P)L,$$

$$K^n(t) = -\frac{1}{\hbar^2}(1 - P)Le^{-\frac{\imath}{\hbar}PLt}PL(1 - P),$$

and

$$I^n(t) = -\frac{\imath}{\hbar}(1 - P)Le^{-\frac{\imath}{\hbar}PLt}P\rho(0).$$

Here we have to pay special attention to the application of the Markovian approximation. The off-diagonal part of the density operator may be slowly varying only in the interaction picture. Otherwise it oscillates with the eigenfrequencies of the unperturbed system. Therefore, in the case of the off-diagonal density operator the Markovian approximation for asymptotic times, under the integral reads as

$$\rho_n(t - t') \Rightarrow e^{\frac{i}{\hbar}L_0(t-t')}\rho_n(t)^I = e^{-\frac{i}{\hbar}L_0 t'}\rho_n(t)\,,$$

followed by

$$\int_0^t dt' \cdots \Rightarrow \int_0^\infty dt' \cdots .$$

Hence, the collision term for the off-diagonal equation looks (to lowest order in L_1) like

$$-\frac{1}{\hbar^2}\int_0^\infty dt'(1-P)L_1 P L_1 e^{-\frac{i}{\hbar}L_0 t'}\rho_n(t)\,.$$

Here we have again used the fact that $PL_0 = L_0 P = 0$. Then, we get

$$\int_0^\infty dt' K_0^n(t')\rho_n(t) = -\frac{1}{\hbar^2}\int_0^\infty dt'\left[H_1,\left[H_1, e^{-\frac{i}{\hbar}H_0 t'}\rho_n(t)e^{-\frac{i}{\hbar}H_0 t'}\right]_d\right]_n.$$

Now, taking the matrix element $< \nu|\cdots|\nu' >$ (with $\nu \neq \nu'$) we have, with the simplifying notation $\quad e^{-\frac{i}{\hbar}L_0 t'}\rho_n(t) \equiv \rho_n^{t'}(t)\,,$ and

$$< \nu|\int_0^\infty dt' K_0^n(t')\rho_n(t)|\nu' >$$

$$= -\frac{1}{\hbar^2}< \nu|H_1|\nu' >\left\{< \nu'|\left[H_1, \rho_n^{t'}(t)\right]|\nu' > - < \nu|\left[H_1, \rho_n^{t'}(t)\right]|\nu >\right\}$$

$$= -\left.< \nu|H_1|\mu >< \mu|\rho_n^{t'}(t)|\nu > + < \nu|\rho_n^{t'}(t)|\mu >< \mu|H_1|\nu >\right\}.$$

If we use also the so called "secular approximation", which consists of omitting the terms that have different free oscillation frequencies as the left-hand side, we are left with

$$\frac{d}{dt}< \nu|\rho|\nu' >|_{coll} = \frac{2i}{\hbar}\frac{1}{E_\nu - E_{\nu'} - i0}|< \nu|H_1|\nu' >|^2 < \nu|\rho|\nu' >\,.$$

If we ignore also the principal value parts (energy corrections), then we get the collision term for the off-diagonal Master equation

$$\frac{d}{dt} < \nu|\rho|\nu' > |_{coll} = -2\pi\delta(E_\nu - E_{\nu'})| < \nu|H_1|\nu' > |^2 < \nu|\rho|\nu' > , \quad (19.7)$$

which seems to describe an exponential decay of the off-diagonal part of the density operator. The extremely formal nature of the result here is again shown by the apparition of the energy conserving delta-function. In our case, it is aggravated by the absence - even in the thermodynamic limit - of any integration over the energy variable.

To conclude, the Master equations derived by the projection operator method do not stay on a firm basis, and their use must be taken with utmost precaution. We will comment on this aspects in the next chapter.

Chapter 20

Van Hove's weak coupling limit and Davies's generalization

Though we have described several derivations of the Master equations, none of them could serve as a proof. They were mere descriptions of the approximations that allowed to obtain - formally though - the desired Markovian equations. A decisive step forward, leading to the understanding of the matter was the paper of van Hove [van Hove (1955)], who tried to give a proof. Although still far from being rigorous from mathematical point of view, he brought in several ideas of fundamental importance.

First of all, he recognized that a Master equation in a closed system can be meaningful only in the thermodynamic limit, therefore, from the beginning he cared for a continuous spectrum. Secondly, he demanded that the interaction posses a certain "diagonal dominance" property. Lastly, after having remarked the scale invariance of the Master equation - based on the use of the "golden rule" - against a simultaneous scaling of the time variable and coupling constant as $t \rightarrow \frac{1}{\lambda^2}t$ and $g \rightarrow \lambda g$, he formulated a clear statement about the limit $\lambda \rightarrow 0$ (weak coupling limit) in which Markovian behavior is to be expected. However, his discussion is restricted to the case of the diagonal Master equation.

After this breakthrough, Davies [Davies (1967)] was able - using van Hove's scaling limit - to perform a rigorous proof of the Master equation for the density operator for a finite system in contact with a thermostat, in the frame of a given model.

Van Hove considers a system with internal interactions, no thermostat, in the weak coupling limit. Let us consider the Hamiltonian

$$H + gV,$$

where V is a weak interaction (g-small), its diagonal part being already included in H, with the ortho-normalized eigenstates $|\nu\rangle$

$$H|\nu\rangle = E_\nu|\nu\rangle; \quad \langle\nu'|\nu\rangle = \delta_{\nu\nu'}.$$

It is supposed that in the thermodynamic limit H has a continuous spectrum characterized by their energy E and some other quantum numbers α.

As we have already discussed it in the previous chapters, the diagonal Master equation for the probability P_ν of a state $|\nu\rangle$ would look as

$$\frac{d}{dt}P_\nu = -\sum_{\nu'}\left(W_{\nu\nu'}P_\nu - W_{\nu'\nu}P_{\nu'}\right).$$

Usually it is admitted that the positive, symmetrical transition rates are given by the "golden rule"

$$W_{\nu\nu'} = 2\pi g^2 \delta(E_\nu - E_{\nu'})\left|\langle\nu|V|\nu'\rangle\right|^2.$$

The presence of the energy conserving δ-function already indicates the meaninglessness of the above equation in a discrete spectrum. Yet we have all the reason to hope that in the thermodynamic limit with a continuous spectrum it will become meaningful. In the mixed ensemble described by the probabilities P_ν the average of an observable A that commutes with the Hamiltonian H, having the eigenvalues a_ν, is given by

$$\bar{A} = \sum_\nu P_\nu a_\nu,$$

which in the thermodynamic limit will go over into

$$\bar{A} \Rightarrow \int dEd\alpha P(E, \alpha)a_{E,\alpha},$$

where the integration occurs over all the quantum numbers and $P(E, \alpha)$ is a probability density. Actually, for this probability density one expects the Master equation to hold

$$\frac{d}{dt}P(E\alpha; t) = -\int dE' \int d\alpha' \{W(E\alpha, E'\alpha')P(E\alpha; t)$$

$$- W(E'\alpha', E\alpha)P(E'\alpha' : t)\}$$

$$\equiv -\int dE' \int d\alpha' A(E\alpha, E'\alpha')P(E'\alpha' : t),$$

with transition rate densities:

$$W(E\alpha, E'\alpha') = 2\pi g^2 \delta(E - E') |\langle E\alpha|V|E'\alpha'\rangle|^2 \equiv \delta(E - E') w_E(\alpha, \alpha').$$

After integrating out the energy variable one gets the Master equation at a fixed energy E

$$\frac{d}{dt}P(E\alpha; t) = -\int d\alpha' \left\{ w_E(\alpha, \alpha')P(E\alpha; t) - w(\alpha', \alpha)P(E\alpha' : t) \right\}$$

$$\equiv -\int d\alpha' A_E(\alpha, \alpha')P(E\alpha' : t).$$

The formal solution is

$$P(t) = e^{-At}P(0),$$

and with the initial condition $P(E\alpha; 0)$, a development in powers of g (actually in powers of $g^2 t$) gives

$$P(E\alpha; t) = P(E\alpha; 0) + t \int d\alpha' w_E(\alpha, \alpha') \left(P(E\alpha'; 0) - P(E\alpha : 0) \right) \quad (20.1)$$

$$+ \frac{1}{2} t^2 \left\{ \left(\int d\alpha' w_E(\alpha, \alpha') \right)^2 P(E\alpha : 0) \right.$$

$$+ \int d\alpha' \left[\int d\alpha'' w_E(\alpha, \alpha'') w_E(\alpha'', \alpha') \right.$$

$$- w_E(\alpha, \alpha') \left(\int d\alpha'' w_E(\alpha, \alpha'') \right.$$

$$\left. \left. \left. + \int d\alpha'' w_E(\alpha', \alpha'') \right) \right] P(E\alpha'; 0) \right\}$$

$$+ \dots .$$

Van Hove makes a systematic comparison of this series with the perturbational series for the probability

$$P_\nu(t) = \sum_{\nu'} |\langle \nu|U|\nu'\rangle|^2 P_{\nu'}(0), \quad (20.2)$$

as it is given by the quantum mechanical evolution with the unitary operator

$$U(t) = e^{-i(H+gV)t}, \qquad (20.3)$$

(here overall $\hbar = 1$), after taking the thermodynamic and his $g \to 0$, $t \to \infty$ $(g^2 t = \tilde{t})$ scaling limit in a mixed ensemble with a diagonal initial condition for the density operator.

We won't reproduce here his original - relatively complicated - reasoning, in which also partial summations of the perturbation series play an important role; we will simply discuss a few terms of the perturbation series in order to illustrate his ideas. The unitary evolution operator $U(t)$ has the well-known series expansion in powers of g

$$e^{iHt}U(t) = 1 + \sum_{n=1}^{\infty}(-ig)^n \int_0^t dt_n \int_0^{t_n} dt_{n-1} \dots \int_0^{t_2} dt_1$$

$$\times e^{iHt_n}Ve^{-iH(t_n-t_{n-1})} \dots Ve^{-iHt_1}. \qquad (20.4)$$

The assumed diagonality of the initial density operator in the basis of the Hamiltonian H corresponds to van Hove's idea about "random phases" only in the initial state.

A term $P_\nu(t)^{(n)}$ of order g^n in Eq. (20.2) is built up of terms of the type

$$\sum_{m=0}^{n} \langle \nu|U^{(n-m)}(t)|\nu'\rangle\langle \nu|U^{(m)}(t)|\nu'\rangle^*, \qquad (20.5)$$

where $U^{(m)}(t)$ is the term of order g^m in Eq. (20.4).

A first order contribution $P_\nu(t)^{(1)}$ to Eq. (20.2) is given by the product of the zeroth order term in the development of the unitary operator $U(t)$, with its first order term $U^{(1)}(t)$

$$-ig\sum_{\nu'}P_{\nu'}(0)\langle \nu|e^{iHt}|\nu'\rangle \int_0^t dt_1 \langle \nu'|e^{iH(t_1-t)}Ve^{-iHt_1}|\nu\rangle + c.c..$$

This term vanishes due to the fact that $\langle \nu|e^{iHt}|\nu'\rangle = \delta_{\nu\nu'}\delta(E-E_0)e^{iE_\nu t}$, since the diagonal part of the perturbation was already included in the unperturbed Hamiltonian H.

There are two second order contributions $P_\nu(t)^{(2)}$. One comes from the product of the first order corrections ($m = 1$ in Eq. (20.5)) while the other one from product of the zeroth order with the second order corrections to $U(t)$ ($m = 0, 2$ in Eq. (20.5)) . The first type is given by

$$g^2 \sum_{\nu'} P_{\nu'}(0) \left| \langle \nu | \int_0^t dt_1 e^{iH(t_1-t)} V e^{-iHt_1} |\nu'\rangle \right|^2$$

$$= \lambda^2 \sum_{\nu'} P_{\nu'}(0) |\langle \nu|V|\nu'\rangle|^2 \left(\frac{\sin\left(\frac{E_\nu - E_{\nu'}}{2}t\right)}{\frac{E_\nu - E_{\nu'}}{2}} \right)^2 .$$

Since

$$\lim_{t\to\infty} \frac{1}{t} \left(\frac{\sin(xt)}{x} \right)^2 = \pi\delta(x) ,$$

one gets for this contribution in the van Hove limit $\lambda \to 0$, $t \to 0$, with $g^2 t = \tilde{t}$ constant

$$2\pi\tilde{t} \sum_{\nu'} P_{\nu'}(0) |\langle \nu|V|\nu'\rangle|^2 \delta(E_\nu - E_{\nu'}) . \tag{20.6}$$

The second type contribution ($n = 2$; $m = 0, 2$ in Eq. (20.5)), is:

$$-\lambda^2 \sum_{\nu'} P_{\nu'}(0) \langle \nu|e^{iHt}|\nu'\rangle\langle \nu'| \int_0^t dt_2 \int_0^{t_2} dt_1 e^{iH(t_2-t)} V e^{-iH(t_2-t_1)} V e^{-iHt_1} |\nu\rangle ,$$

plus its complex conjugate or

$$-2g^2 P_\nu(0) \sum_{\nu'} |\langle \nu|V|\nu'\rangle|^2 \Im \left(\int_0^t dt_2 \int_0^{t_2} dt_1 e^{iE_\nu(t_2-t_1)} e^{iE_{\nu'}(t_1-t_2)} \right) ;$$

and since

$$\int_0^t dt_2 \int_0^{t_2} dt_1 e^{iE(t_2-t_1)} e^{iE'(t_1-t_2)} = \frac{1 - e^{-i(E-E')t} + i(E-E')t}{(E-E')^2} ,$$

in the van Hove limit $g \to 0$, $t \to \infty$, ($g^2 t = \tilde{t}$ constant) survives only

$$2\tilde{t}P_\nu(0)\sum_{\nu'}|\langle\nu|V|\nu'\rangle|^2\Im\frac{1}{E_\nu - E_{\nu'} + i0}\,.$$

By considering this limit, however, we had to replace E_ν by $E_\nu + i0$ in order to give a mathematical meaning to the limiting expression. Finally, we get for the second type of contribution

$$-2\pi\tilde{t}P_\nu(0)\sum_{\nu'}|\langle\nu|V|\nu'\rangle|^2\delta(E_\nu - E_{\nu'})\,. \tag{20.7}$$

Again both results Eq. (20.6) and Eq. (20.7) contain the energy conserving delta function. Therefore, in order to obtain meaningful results, one had to perform first the thermodynamic limit, and only after that van Hove's scaling limit

$$\lim_{t\to\infty,g\to 0,(g^2 t=\tilde{t})}\quad\lim_{\Omega\to\infty,N\to\infty,(\frac{N}{\Omega}=n)}\,.$$

Indeed, in this case, the two lowest order terms we obtained from the quantum-mechanical evolution coincide with the second order corrections in the solution of the Master equation Eq.(20.1).

It can be shown that third order terms in g do not survive in the van Hove limit.

The discussion of the fourth order term requires already the implementation of the "diagonal dominance" principle of van Hove. (We will discuss its justification later.) We illustrate its use on the example of the diagonal term that comes from the product of a zeroth order term in the unitary operator U with its fourth order term. The terms ($n = 4$; $m = 0, 4$ in Eq. (20.5)) are

$$g^4 P_\nu(0)\int_0^t dt_4\int_0^{t_4}dt_3\int_0^{t_3}dt_2\int_0^{t_2}dt_1$$
$$\times\,\langle\nu|e^{iHt_4}Ve^{-iH(t_4-t_3)}Ve^{-iH(t_3-t_2)}Ve^{-iH(t_2-t_1)}Ve^{-iHt_1}|\nu\rangle\,,$$

together with its complex conjugate. Now let us split the product of the four operators V in two pieces

$$\langle\nu|e^{iHt_4}Ve^{-iH(t_4-t_3)}Ve^{-iH(t_3-t_2)}Ve^{-iH(t_2-t_1)}Ve^{-iHt_1}|\nu\rangle$$
$$=\sum_{\nu'}e^{iE_\nu(t_4-t_1)}\langle\nu|Ve^{-iH(t_4-t_3)}V|\nu'\rangle e^{-iE_{\nu'}(t_3-t_2)}\langle\nu'|Ve^{-iH(t_2-t_1)}V|\nu\rangle\,.$$

The requirement of the "diagonal dominance" property of the perturbation, introduced by van Hove, states that in the thermodynamic limit the diagonal elements

$$\langle \nu | V e^{-\imath H(t_4-t_3)} V | \nu \rangle$$

are dominating in the summation. In the continuum limit this amounts to a singular δ-function behavior. Then, one may retain only

$$g^4 P_\nu(0) \int_0^t dt_4 \int_0^{t_4} dt_3 \int_0^{t_3} dt_2 \int_0^{t_2} dt_1$$

$$\times \langle \nu | e^{\imath H t_4} V e^{-\imath H(t_4-t_3)} V e^{-\imath H(t_3-t_2)} | \nu \rangle \langle \nu | V e^{-\imath H(t_2-t_1)} V e^{-\imath H t_1} | \nu \rangle \,,$$

or

$$g^4 P_\nu(0) \int_0^t dt_4 \int_0^{t_4} dt_3 \int_0^{t_3} dt_2 \int_0^{t_2} dt_1$$

$$\times \langle \nu | V e^{-\imath(H-E_\nu)(t_4-t_3)} V | \nu \rangle \langle \nu | V e^{-\imath(H-E_\nu)(t_2-t_1)} V | \nu \rangle \,.$$

But the integral

$$\int_0^t dt_4 \int_0^{t_4} dt_3 \int_0^{t_3} dt_2 \int_0^{t_2} dt_1 e^{-\imath E(t_4-t_3)} e^{-\imath E'(t_2-t_1)}$$

has for $t \to \infty$ a leading term

$$\frac{t^2}{2(E-\imath 0)(E'+\imath 0)} \,.$$

This is determined by the fact that one has four time integrations, while the integrand depends only on two independent time variables. (Here again the infinitesimal imaginary parts were introduced in order to define the improper limit.) Therefore, we get for the discussed fourth order term

$$\tilde{t}^2 P_\nu(0) \left| \sum_{\nu'} |\langle \nu | V | \nu' \rangle|^2 \frac{1}{E_\nu - E_{\nu'} + \imath 0} \right|^2 \,.$$

Still, we have to take into account that an identical term comes also from the diagonal term in the product of the two second order corrections to the unitary operator U ($n = 4$; $m = 2$ in Eq.(20.5)), however, in those contributions one does not need to use the "diagonal dominance". Therefore the total diagonal term is

$$2\tilde{t}^2 P_\nu(0) \left| \sum_{\nu'} |\langle \nu|V|\nu'\rangle|^2 \frac{1}{E_\nu - E_{\nu'} + i0} \right|^2 ,$$

or

$$\frac{1}{2}\tilde{t}^2 P_\nu(0) \left(\sum_{\nu'} 2\pi |\langle \nu|V|\nu'\rangle|^2 \delta(E_\nu - E_{\nu'}) \right)^2 ,$$

plus the square of the principal value term

$$\frac{1}{2}\tilde{t}^2 P_\nu(0) \left(\sum_{\nu'} (2\pi |\langle \nu|V|\nu'\rangle|^2 P \frac{1}{E_\nu - E_{\nu'}} \right)^2 .$$

Up to this square of the principal value terms, this is similar to the diagonal fourth order term in the solution of the Master equation. As we well know, such principal value terms usually represent energy corrections. Indeed, in his original paper, van Hove gets rid of the principal value terms by first performing a partial summation of the series. Then, the the principal value terms may be summed up to energy corrections in the energy conserving δ - functions, and finally disappear in the weak coupling limit. His analysis shows systematically that in the assumed thermodynamic and scaling limits as well as the repeated use of the "diagonal dominance", the whole series for the probability gives rise to the series of the solution of the Markovian Master equation.

As we have seen, the "diagonal dominance" is essential in the derivation. The argumentation of van Hove - for its justification - can be illustrated on the example of the particle-particle interaction

$$V = \frac{1}{2} \int d\vec{x} \int d\vec{x}' \psi^+(\vec{x})\psi^+(\vec{x}')v(|\vec{x} - \vec{x'}|)\psi(\vec{x'})\psi(\vec{x}) ,$$

or in the Fock-space of plane wave-states

$$V = \frac{1}{2} \sum_{\vec{k}, \vec{k}', \vec{q}} \tilde{v}(q) a_{\vec{k}}^+ a_{\vec{k}'}^+ a_{\vec{k}'+\vec{q}} a_{\vec{k}-\vec{q}}.$$

Here the conservation of the total momentum is enforced in a finite box through periodical boundary conditions. Let $n_{\vec{k}} = 0, 1$ be the fermion occupation number of a state of wave-vector $\vec{k}$.

Now let us take a matrix element of the square of V between two Fock states having certain sets of occupation numbers n and n' and let us insert a complete set of Fock states n'' between the two interaction operators

$$\langle n | V^2 | n' \rangle = \sum_{n''} \langle n | V | n'' \rangle \langle n'' | V | n' \rangle.$$

From the structure of the interaction V it follows that the state n'' may differ from the state n just by the "scattering" of two particles and in the same time it may differ from the state n' also only by the "scattering" of two particles. If the states n and n' however coincide, then there is only a single condition imposed on the state n''. Therefore, due to the fact that more intermediate states contribute to the diagonal matrix elements of V^2 and to the fact that their contribution is positive (due to the hermiticity of V), and their number increases unlimited in the thermodynamic limit, it is reasonable to think that the diagonal terms dominate in a singular manner.

Although it is very systematic, the perturbational analysis of van Hove is not a rigorous proof. Resummation of series, after performing limits on it, is - from a mathematical point of view - not at all legitimate. Nevertheless, the elements of his reasoning seem to be fundamental. We call the attention that van Hove's discussion says nothing about the evolution of non-diagonal matrix elements (see also the comments at the end of the previous Chapter).

A rigorous mathematical proof of the diagonal and non-diagonal Master equation was given later by Davies [Davies (1967)] (see also [Gorini, Frigerio, Verry, Kossakowski and Sudarshan (1978)] for a nice presentation), using some of van Hove's ideas. He considers, however, a finite system (A) in contact to a thermostat (B). This problem is much better defined mathematically.

Davies starts from the exact equation Eq.(17.2) for the density matrix of the subsystem A in the interaction picture we already derived, while

discussing reservoir theory.   The assumptions about the initial state
Eq.(17.1) as well as the properties of the interaction Hamiltonian Eqs.(17.3)
and (17.5) are the same as there. However Davies does not make any further ad hoc assumptions, but proves that in van Hove's scaling limit $g \to 0$,
$t \to \infty$ ($g^2 t =$ constant) the (diagonal and non-diagonal) density operator
of the subsystem A tends in a mathematically unambiguous sense to the
solution of Eq.(17.12), formerly obtained by using several uncontrolled assumptions within the frame of the reservoir theory. Using the notations of
Chapter 17

$$\lim_{t \to \infty, g \to 0, \, g^2 t = const.} e^{\frac{i}{\hbar} H_A t} Tr_B \rho(t) e^{-\frac{i}{\hbar} H_A t} = e^{-K^{\natural} t} \rho_A(0). \qquad (20.8)$$

According to Davies, this statement has to be understood in the sense of
the vanishing of the norm of the difference of the two sides after taking the
limits.

Of course, van Hove's scaling limit is a weak couplings and long times
limit. On short time scales, as we have seen on the example of solvable
classical models, and as we shall discuss later within the frame of quantum
kinetics, the behavior is far from being irreversible. On the other hand,
the influence of the interaction may be reduced to simple transitions (scattering) between unperturbed states only in the case of a weak coupling.
The same holds for the interaction with a thermostat. It gives rise to a
macro-canonical distribution in terms of the unperturbed system A only
if the interaction with the thermostat can be ignored in the description
of the states themselves, and this is the approach of the macroscopical
thermodynamics.

Chapter 21

Bose-Einstein condensation in real time

The property of bosons to undergo a specific phase transition - condensation on the lowest level - was predicted many years ago by Einstein. However, until recently there has been no experimental evidence of the Bose-Einstein condensation (BEC). The observation of the fascinating BEC of atoms in magnetic traps in 1999 [Anderson et al. (1995); Davis et al. (1995)] raised again the interest for this quantum mechanical phenomenon on macroscopic scale. There is a sustained experimental effort to put into evidence BEC also for excitons in semiconductors.

One of the most important characteristics of this phase transition is that, due to its macroscopical nature, a spontaneous symmetry breaking occurs, which is characterized by an order parameter; namely, a non-vanishing average of the second quantized wave function of bosons $\langle \psi(x) \rangle$. This may be demonstrated experimentally by macroscopic matter interference phenomena.

Here we are interested in a theoretical understanding of BEC as it occurs in real time, i.e. not as mere an equilibrium problem. This description may be looked upon as a model for the description of phase transitions with spontaneous symmetry breaking in real time. Our attention is devoted to the phenomenon of spontaneous symmetry breaking as a characteristic non-linear dynamic phenomenon [Bányai, Gartner, Schmitt, and Haug (2000)]. It is important to understand in this context the role of the thermodynamic limit, as well as its relevance for systems with large, but still finite volumes.

We are going to deal here only with the case of bosons interacting with a thermostat, but not interacting within themselves. This case is clear cut, and may be interpreted within the frame of a Markovian description of non-equilibrium.

The interaction between the bosons would introduce essential modifications in the picture, even within the simplest self-consistent scheme. Indeed, while in the case of the equilibrium theory one may speak of quasiparticles - in terms of which the discussion may structured - in non-equilibrium all the parameters (averages) will be time dependent, and no such simplification is available. In extremis, one may imagine an adiabatic theory with slowly time-dependent quasiparticle parameters, however, its use would be very limited, partly due to the arbitrariness in the Berry phase, but also to the real rate of the process. In this case, a more appropriate approach is the quantum kinetic one. We need such an approach not only because the rapidity of the occurring processes, but also because within any self-consistent approximation the Hamiltonian is time-dependent. The emerging quantum -kinetic equations, however, are not well understood mathematically, although numerical solutions are available with encouraging results [Schmitt, Tran Thoai, Bányai, Gartner, and Haug (2001)]. Later we will discuss such an approach to other phenomena related to the optics of semiconductors interacting with ultra-short light pulses.

To fix the frame, we describe first the BEC of a free gas of bosons as an equilibrium phase transition. Then we derive the equation for the time evolution of the order parameter in a system of non-interacting bosons, which has to be considered together with the rate equation for the average occupation numbers. We consider here only collision terms appropriate for the interaction with a thermal bath. The mathematical properties of the equations are discussed in detail for the case of free bosons with special emphasis on the peculiarities of the thermodynamic limit. Last but not least, we are going to present a much better insight into the aspects of the spontaneous symmetry breaking, and the way it is related to the volume of the system on the example of an exactly solvable model.

21.1 BEC of the free Bose gas as an equilibrium phase transition

Let us consider free bosons of mass m in a cube of volume $V = L^3$ with periodic boundary conditions. The one-particle energy is $e_{\vec{k}} = \frac{\hbar^2 \vec{k}^2}{2m}$ with discrete wave-vector $(k_x, k_y, k_z = 0, \pm\frac{2\pi}{L}, \pm 2\frac{\pi}{L}, \dots)$. We want to calculate the grand-canonical partition function (statistical sum)

$$Z = Tr\{e^{-\beta(H-\mu N)}\},$$

and thereafter the relevant thermodynamic entities within Bogolyubov's quasi-average formulation of phase transitions with spontaneous symmetry breaking (of the U(1) symmetry). This symmetry (against one dimensional phase transformations) emerges from the conservation of the total particle number. According to Bogolyubov's quasi-averages concept, one has to introduce a symmetry breaking term in the Hamiltonian, characterized by an infinitesimal parameter λ:

$$H - \mu N = \sum_{\vec{k}} (e_{\vec{k}} - \mu) a_{\vec{k}}^+ a_{\vec{k}} + \lambda^* \sqrt{V} a_0 + \lambda \sqrt{V} a_0^+ ,$$

and perform first the thermodynamic limit, and only thereafter the $\lambda \to 0$ limit. For $\lambda = 0$ of course $< a_0 > = 0$ in any finite volume.

Let us bring this Hamiltonian to a diagonal quadratic form by the introduction of shifted annihilation operators (satisfying the same bosonic commutation rules)

$$A_0 = a_0 + \frac{\lambda \sqrt{V}}{\mu}.$$

Then

$$H - \mu N = \sum_{\vec{k} \neq 0} (e_{\vec{k}} - \mu) a_{\vec{k}}^+ a_{\vec{k}} - \mu A_0^+ A_0 + \frac{|\lambda|^2 V}{\mu} ,$$

giving rise to the free energy

$$F = \frac{1}{\beta V} \sum_{\vec{k}} \ln \left\{ 1 - e^{-\beta(e_{\vec{k}} - \mu)} \right\} + \frac{|\lambda|^2}{\mu} ,$$

and the average particle density:

$$\frac{< N >}{V} = \frac{1}{V} \sum_{\vec{k}} \frac{1}{e^{\beta(e_{\vec{k}} - \mu)} - 1} + \frac{|\lambda|^2}{\mu^2} . \tag{21.1}$$

For $\lambda \neq 0$ this last equation may be satisfied for any density with a strictly negative chemical potential $\mu < 0$, which would not be the case for $\lambda = 0$. In the thermodynamic limit the Riemann sum goes over to an integral

$$\frac{1}{V}\sum_{\vec{k}} \rightarrow \frac{1}{(2\pi)^3}\int d\vec{k}\,,$$

while the density ($n \equiv \frac{<N>}{V}$) is kept constant. Then, the relation between the density n and the chemical potential μ is defined by the transcendent equation

$$n = \frac{|\lambda|^2}{\mu^2} + \frac{1}{(2\pi)^3}\int d\vec{k}\frac{1}{e^{\beta(e(\vec{k})-\mu)}-1}\,. \tag{21.2}$$

On the other hand, for any allowed negative chemical potential $\mu \le 0$ one has the inequality

$$\frac{1}{(2\pi)^3}\int d\vec{k}\frac{1}{e^{\beta(e(\vec{k})-\mu)}-1} \le \frac{1}{(2\pi)^3}\int d\vec{k}\frac{1}{e^{\beta e(\vec{k})}-1} = 4.63\pi\left(\frac{2m}{\beta\hbar^2}\right)^{\frac{3}{2}} \equiv n_c\,.$$

While for densities below the critical one n_c, one gets a solution with a finite strictly negative chemical potential $\mu < 0$ (implying also a vanishing order parameter), above this critical density the first term in Eq. (21.2) has to compensate the difference. This situation means that for $\lambda \to 0$ the chemical potential has to go to zero as $\mu \to \frac{\lambda}{\sqrt{n_0}}$ in order to make up for the difference $n_0 = n - n_c$. In the same time the order parameter also will survive as

$$P \equiv \lim_{\lambda \to 0}\lim_{V \to \infty}\frac{<a_0>}{\sqrt{V}} = \sqrt{n_0}e^{i\phi}\,,$$

where ϕ is the phase of λ ($\lambda = |\lambda|e^{i\phi}$).

Another way to see the existence of the critical point - without introduction of the infinitesimal symmetry breaking - is through the inspection of Eq. (21.1) at $\lambda \equiv 0$. *Indeed, in order to get a solution for $n > n_c$ the chemical potential must go to zero with the volume as $\mu \sim -\frac{1}{n_0\beta V}$.* This implies, however, that the discrete sum does not go over to a Riemann integral, but one has to separate the $\vec{k} = 0$ term, whose contribution is not any more infinitesimal.

The order parameter $\frac{<a_0>}{\sqrt{V}}$ plays here the same role as the magnetization in the theory of ferromagnetic transitions. The spontaneously broken symmetry is the particle number conservation (equivalent to a phase invariance), while in the case of a ferromagnetic transition the rotation symmetry is broken spontaneously.

21.2 Markovian real-time description of non-interacting bosons in contact with a thermal bath

Let us consider non-interacting bosons described by some quantum numbers α, interacting with a thermal bath (phonons) characterized by a quantum number q by the Hamiltonian

$$H = \sum_\alpha e_\alpha a_\alpha^+ a_\alpha + \sum_q \hbar\omega_q b_q^+ b_q + \sum_{\alpha,\alpha',q} g_{q,\alpha\alpha'} a_\alpha^+ a_{\alpha'} (b_q + b_{-q}^+) ,$$

with $g_{q,\alpha\alpha'} = g_{-q,\alpha'\alpha}^*$ for hermiticity.

With the same method we used earlier for fermions, one can derive in the Markov limit a rate equation for the occupation numbers of bosons

$$\frac{\partial}{\partial t} \langle a_\alpha^+ a_\alpha \rangle_t = \sum_{\alpha'} \left\{ W_{\alpha'\alpha} \langle a_{\alpha'}^+ a_{\alpha'} \rangle_t (1 + \langle a_\alpha^+ a_\alpha \rangle_t) - (\alpha \rightleftharpoons \alpha') \right\} , \quad (21.3)$$

where $W_{\alpha\alpha'}$ are the "golden rule" transition rates due to the interaction with the thermal bath

$$W_{\alpha\alpha'} = \frac{2\pi}{\hbar} \sum_{q,\pm} |g_{q,\alpha\alpha'}|^2 \pi\delta(e_\alpha \mp \hbar\omega_q) \mathcal{N}_q^\pm ,$$

with

$$\mathcal{N}_q^\pm \equiv \langle b_q^+ b_q \rangle_0 + \frac{1}{2} \pm \frac{1}{2} .$$

These transition rates satisfy the "detailed balance" relation

$$W_{\alpha\alpha'} = W_{\alpha'\alpha} e^{\beta(e_\alpha - e_{\alpha'})} .$$

The only difference to the fermionic case is the sign in the final state factor corresponding to the Bose statistics.

In the bosonic case, we have to derive also an equation describing the time-evolution of the order parameter (symmetry breaking), which is the average of the annihilation operator of the lowest state $\langle a_0 \rangle$. By commuting the annihilation operator a_0 with H and performing the averaging, one gets the equation of motion

$$i\hbar \frac{\partial}{\partial t} \langle a_0 \rangle_t = -\sum_{\alpha, q} g_{q,0\,\alpha} \langle a_\alpha \left(b_q + b^+_{-q} \right) \rangle_t . \tag{21.4}$$

Furthermore, one considers the equation of motion for the new average

$$\left[i\hbar \frac{\partial}{\partial t} - (e_\alpha - \hbar \omega_q) \right] \langle a_\alpha b^+_q \rangle_t$$

$$= \sum_{\alpha', \alpha'', q'} g_{q', \alpha', \alpha''} \left(\langle a^+_{\alpha'} a_{\alpha''} a_\alpha \rangle_t \delta_{qq'} - \delta_{\alpha, \alpha'} \langle a_{\alpha''} b^+_q (b_{q'} + b^+_{-q'}) \rangle_t \right) ,$$

and a similar conjugate equation.

We perform now a factorization of the averages in the last equation, retaining only the basic variables $\langle a^+_\alpha a_\alpha \rangle$, $\langle b^+_q b_q \rangle$, $\langle a_0 \rangle$ and $\langle a^+_0 \rangle$. Moreover, we replace the average phonon occupation number with its equilibrium value $\langle b^+_q b_q \rangle^0 = \frac{1}{e^{\beta \hbar \omega_q} - 1}$, i.e. we take

$$\langle a^+_{\alpha'} a_{\alpha''} a_\alpha \rangle_t \approx \langle a^+_\alpha a_\alpha \rangle_t \delta_{\alpha' \alpha} \delta_{\alpha'', 0} \langle a_0 \rangle_t,$$

$$\langle a_{\alpha''} b^+_q b_{q'} \rangle_t \approx \delta_{\alpha'', 0} \delta_{q, q'} \langle a_0 \rangle_t \langle b^+_q b_q \rangle^0,$$

and therefore

$$\left[i\hbar \frac{\partial}{\partial t} - (e_\alpha - \hbar \omega_q) \right] \langle a_\alpha b^+_q \rangle_t = \langle a_0 \rangle_t g_{q\alpha 0} \left(\langle a^+_\alpha a_\alpha \rangle_t - \langle b^+_q b_q \rangle^0 \right) .$$

This can be written also in a more convenient form suggesting the transitions

$$\left[i\hbar \frac{\partial}{\partial t} - (e_\alpha - \hbar \omega_q) \right] \langle a_\alpha b^+_q \rangle_t = \langle a_0 \rangle_t g_{q\alpha 0} \left(\langle a^+_\alpha a_\alpha \rangle_t \left(\langle b^+_q b_q \rangle^0 + 1 \right) \right.$$

$$\left. - \left(\langle a^+_\alpha a_\alpha \rangle_t + 1 \right) \langle b^+_q b_q \rangle^0 \right) .$$

A formal integration with vanishing initial condition for $\langle a_\alpha b^+_q \rangle$ at $t = 0$

gives

$$\langle a_\alpha b_q^+ \rangle_t = \frac{1}{i\hbar} g_{q\alpha 0} \int_0^t dt' e^{-\frac{i}{\hbar}(e_\alpha - \hbar\omega_q)(t-t')} \langle a_0 \rangle_{t'} \left[\langle a_\alpha^+ a_\alpha \rangle_{t'} \left(\langle b_q^+ b_q \rangle^0 + 1 \right) \right.$$
$$\left. - \left(\langle a_\alpha^+ a_\alpha \rangle_{t'} + 1 \right) \langle b_q^+ b_q \rangle^0 \right] .$$

A similar result holds also for $< a_\alpha b_{-q} >_t$. Inserting these results in Eq. (21.4) we get

$$\frac{\partial}{\partial t} \langle a_0 \rangle_t = \frac{1}{\hbar^2} \sum_{\alpha, q} |g_{q\alpha 0}|^2 \int_0^t dt' \langle a_0 \rangle_{t'}$$

$$\times \left\{ e^{-\frac{i}{\hbar}(e_\alpha - \hbar\omega_q)(t-t')} \left[\langle a_\alpha^+ a_\alpha \rangle_{t'} (\langle b_q^+ b_q \rangle_0 + 1) - (\langle a_\alpha^+ a_\alpha \rangle_{t'} + 1)\langle b_\alpha^+ b_\alpha \rangle_0 \right] \right.$$

$$\left. + e^{-\frac{i}{\hbar}(e_\alpha + \hbar\omega_{-q})(t-t')} \left[\langle a_\alpha^+ a_\alpha \rangle_{t'} \langle b_q^+ b_q \rangle_0 - (\langle a_\alpha^+ a_\alpha \rangle_{t'} + 1)(\langle b_\alpha^+ b_\alpha \rangle_0 + 1) \right] \right\} .$$

In the next step one considers the variables as slowly varying in time, and pulls them outside the time integral. Moreover, by pushing the time integration to infinity, one gets the Markov approximation:

$$\frac{\partial}{\partial t} \langle a_0 \rangle_t = \frac{1}{\hbar} \sum_{\alpha, q, \pm} |g_{q\alpha 0}|^2 \tag{21.5}$$

$$\times \left\{ \left(\pi \delta(e_\alpha \mp \hbar\omega_q) - i P \frac{1}{e_\alpha \mp \hbar\omega_q} \right) \left[\langle a_\alpha^+ a_\alpha \rangle_t \mathcal{N}_q^\pm - (\langle a_\alpha^+ a_\alpha \rangle_t + 1) \mathcal{N}_q^\mp \right] \right\} ,$$

with $\mathcal{N}_q^\pm \equiv \frac{1}{e^{\beta\hbar\omega_q} - 1} + \frac{1}{2} \pm \frac{1}{2}$.

It is easy to see that the equation for the modulus square of the order parameter can be written down also in terms of the transition rates alone, like the rate equation for the average occupation numbers

$$\frac{\partial}{\partial t} |\langle a_0 \rangle_t|^2 = |\langle a_0 \rangle_t|^2 \sum_{\alpha \neq 0} \left(\langle a_\alpha^+ a_\alpha \rangle_t W_{\alpha 0} - (1 + \langle a_\alpha^+ a_\alpha \rangle_t) W_{0\alpha} \right) . \tag{21.6}$$

The principal value terms from Eq.(21.5) give rise to a slow variation of the phase of the order parameter due to the small energy correction of the ground state energy produced by the interaction with the phonons. This correction is not essential, and therefore we will concentrate on equations Eq. (21.3)and Eq. (21.6) as the basis for a real-time description of BEC of non-interacting particles. Their formal stationary solution is the Bose distribution for the population and a vanishing order parameter. This solution is also a stable one, if the transition rates satisfy the already discussed connectivity property (see Chapter 8). As we shall see later, the thermodynamic limit, like in the previous treatment of the equilibrium, introduces essential deviations from this formal argument.

21.3 BEC in real time of the free Bose gas

Now, let us consider the definite case of free bosons in contact to a thermal bath. The rate equation for the population is

$$\frac{\partial}{\partial t}\langle a_{\vec{k}}^{+}a_{\vec{k}}\rangle = -\frac{1}{V}\sum_{\vec{k}'}\left\{W_{\vec{k}\vec{k}'}\langle a_{\vec{k}}^{+}a_{\vec{k}}\rangle(1+\langle a_{\vec{k}'}^{+}a_{\vec{k}'}\rangle) - (\vec{k}\rightleftharpoons\vec{k}')\right\}, \qquad (21.7)$$

with the "detailed balance":

$$W_{\vec{k}\vec{k}'} = W_{\vec{k}'\vec{k}}\,e^{\beta(e_{\vec{k}}-e_{\vec{k}'})}.$$

Here we put into evidence the important volume factor $\frac{1}{V}$ that comes from the plane wave character of the initial and final states. In this sense, the definition of the transition rates differs from the previous one. This equation is decoupled from the equation of the order parameter, and their properties may be discussed separately.

The rate equation Eq.(21.7)- in any finite volume V - conserves the total average number of bosons

$$\frac{\partial}{\partial t}\sum_{\vec{k}}\langle a_{\vec{k}}^{+}a_{\vec{k}}\rangle = 0; \qquad\qquad \langle a_{\vec{k}}^{+}a_{\vec{k}}\rangle \geq 0\,,$$

and its stationary solution (*stable fixed point*) is the Bose distribution

$$\langle a_{\vec{k}}^{+}a_{\vec{k}}\rangle^{0} \equiv \frac{1}{e^{\beta(e_{\vec{k}}-\mu)}-1}\,,$$

with a chemical potential μ below the energy of the lowest state, which here is 0.

However, before performing the thermodynamic limit, we must take into account the delicate mathematical aspects we have seen by the discussion of the equilibrium treatment of the ideal Bose gas, emerging from the fact that the population of the lowest lying ($\vec{k}=0$) state might become extensive.

Therefore, we must distinguish two basic population variables. The normal population

$$f_{\vec{k}}(t) \equiv \langle a_{\vec{k}}(t)^+ a_{\vec{k}}(t) \rangle \qquad (\vec{k} \neq 0) \,,$$

and the condensate

$$n_0(t) \equiv \frac{1}{V} \langle a_0(t)^+ a_0(t) \rangle \,.$$

Separating the non-Riemannian (non-infinitesimal) terms in Eq.(21.7) one gets the coupled equations [Bányai, Gartner, Schmitt, and Haug (2000)]

$$\frac{\partial}{\partial t} f_{\vec{k}} = -\frac{1}{V} \sum_{\vec{k}'} \left\{ W_{\vec{k}\vec{k}'} f_{\vec{k}} (1 + f_{\vec{k}'}) - (\vec{k} \rightleftharpoons \vec{k}') \right\} \tag{21.8}$$

$$- \left[W_{\vec{k}0} f_{\vec{k}} - W_{0\vec{k}} (1 + f_{\vec{k}}) \right] n_0 \,, \tag{21.9}$$

$$\frac{\partial}{\partial t} n_0 = \sum_{\vec{k}} \left[W_{\vec{k}0} f_{\vec{k}} - W_{0\vec{k}} (1 + f_{\vec{k}}) \right] n_0 \,. \tag{21.10}$$

Here we already neglected terms of order $\frac{1}{V}$. Then, performing the thermodynamic limit, i.e. going over to integrals, we get the coupled equations

$$\frac{\partial}{\partial t} f(\vec{k}, t) = -\frac{1}{(2\pi)^3} \int d\vec{k} \left\{ W(\vec{k}, \vec{k}') f(\vec{k}, t)(1 + f(\vec{k}', t)) - (\vec{k} \rightleftharpoons \vec{k}') \right\}$$

$$- \left[W(\vec{k}, 0) f(\vec{k}, t) - W(0, \vec{k})(1 + f(\vec{k}, t)) \right] n_0(t) \,, \tag{21.11}$$

$$\frac{\partial}{\partial t} n_0(t) = \frac{1}{(2\pi)^3} \int d\vec{k} \left[W(\vec{k}, 0) f(\vec{k}, t) - W(0, \vec{k})(1 + f(\vec{k}, t)) \right] n_0(t) \,.$$

These equations also conserve the total average particle density

$$n = n_0(t) + \frac{1}{(2\pi)^3} \int d\vec{k} f(\vec{k}, t) \,.$$

The non-linear equations Eq.(21.11) have two attractive fixed points :
1) a "normal" one

$$f(\vec{k}) = \frac{1}{e^{\beta(e(\vec{k}) - \mu)} - 1} \qquad (\mu < 0); \qquad n_0 = 0 \,,$$

whose basin of attraction contains all the initial conditions with the total density below the critical one n_c, discussed by the equilibrium theory and

2) a "condensed" one

$$f(\vec{k}) = \frac{1}{e^{\beta e(\vec{k})} - 1}; \qquad n_0 \neq 0,$$

whose basin of attraction contains all the initial conditions with the total density above the critical one n_c and simultaneously having any non-vanishing initial macroscopic condensate $n_0(0)$. For the detailed proof of these statements see Appendix A.

In other words, with any over-critical initial total density and even an infinitesimally small initial condensate, one gets asymptotically a complete Bose-Einstein condensation. In order to produce BEC we need only a "seed". As a matter of fact, this "seed" might be produced by fluctuations, which are not included in our description.

A peculiar situation occurs by an over-critical total density in the absence of a condensate "seed". It can be shown [Bányai, Gartner, Schmitt, and Haug (2000)] that in this case the rate equation for $f(\vec{k})$ has no asymptotic solution as an ordinary function. It develops a δ-singularity around $k = 0$.

In deriving the above equations, we have implicitly admitted that $\langle a_0^+(t)a_0(t)\rangle = |\langle a_0(t)\rangle|^2$ (to leading order in $\frac{1}{V}$). By comparing the second of the equations Eq.(21.11) with the equation Eq.(21.6) we derived earlier for the modulus square of the order parameter, we may see that the equations are identical. Therefore, if the relation holds at $t = 0$ ($\langle a_0^+(0)a_0(0)\rangle = |\langle a_0(0)\rangle|^2$), then it holds for any later instant.

One might extend the above picture to include the interaction between the bosons in an adiabatic, self-consistent manner. This may be done in a way similar to the description of the hopping of charged particles. There, the basic assumption was to replace the energies in the transition rates by the self-consistent time-dependent energies emerging form the current charge distribution. In the case of the Bose-Einstein condensation the simplest modification of the theory along these lines is to replace within the transition rates the kinetic energies of the bosons by that of quasiparticles having the time-dependent Bogolyubov spectrum $\epsilon(k,t) = \sqrt{2w|P(t)|^2 \frac{\hbar^2 q^2}{2m} + (\frac{\hbar^2 q^2}{2m})^2}$. The time-dependence here comes through the self-consistent, time-dependent order parameter $P(t)$.

21.4 Spontaneous symmetry breaking in a finite volume. A solvable model of BEC

Here we consider as an example, a solvable model describing Bose-Einstein condensation in real time, with spontaneous symmetry breaking [Bányai and Gartner (2002)]. The system of non-interacting bosons has a nondegenerate one-particle ground-state (particle energy ϵ_0) and an excited state (particle energy ϵ_1) taken to be macroscopically degenerate (i.e. with the degeneracy $V n_1$, proportional to the volume V).

Fig. 21.1 Lowest level non-degenerate, higher level macroscopically degenerate.

The bosons are interacting with a thermostat in equilibrium at the inverse temperature β. The rate equations for the average occupation numbers $f_i \equiv \langle a_i^+ a_i \rangle$ $(i = 0, 1)$ are

$$\frac{\partial}{\partial t} f_1 = -\frac{w}{V} \left[f_1(1 + f_0)e^{-\beta\epsilon_0} - f_0(1 + f_1)e^{-\beta\epsilon_1} \right], \qquad (21.12)$$

$$\frac{\partial}{\partial t} f_0 = -n_1 w \left[f_0(1 + f_1)e^{-\beta\epsilon_1} - f_1(1 + f_0)e^{-\beta\epsilon_0} \right].$$

The presence of the nonlinear terms (final state factors) is typical for the many-body aspect.

The transition rates are taken so as to satisfy the detailed balance, and their volume dependence is taken in analogy to those of free bosons interacting with a thermal bath. Here we admitted implicitly that the occupation numbers of all the degenerate states are identical. This explains the absence of the $\frac{1}{V}$ factor in the second of these equations.

Besides the average occupation numbers it is important to consider also the equation for the order parameter $p \equiv \frac{1}{V}\langle a_0 \rangle$. According to the previous discussion we consider here only the equation for the modulus square of the order parameter $|p|^2$ (see Eq.(21.6))

$$\frac{\partial}{\partial t} |p(t)|^2 = - |p(t)|^2 \, n_1 w \left[(1 + f_1) e^{-\beta \epsilon_1} - f_1 e^{-\beta \epsilon_0} \right] . \qquad (21.13)$$

The already discussed properties hold: the equations conserve the average particle density $n_{tot} \equiv n_1 f_1 + \frac{1}{V} f_0$:

$$\frac{\partial}{\partial t} n_{tot} = 0 , \qquad (21.14)$$

while the equilibrium solution for the occupation numbers is the Bose distribution

$$f_{0,1}^0 = \frac{1}{e^{\beta(\epsilon_{0,1} - \mu)} - 1} ,$$

and a vanishing order parameter

$$p^0 = 0 ,$$

for any finite volume V.

However, since the chemical potential has an upper bound : $\mu \leq \epsilon_0$, the particle density in the state 1 is bounded by $\frac{1}{e^{\beta(\epsilon_0 - \epsilon_1)} - 1}$. Therefore, above a critical density at fixed temperature (or below a critical temperature at fixed n_{tot}) the equilibrium occupation of the state 0 has to increase with the volume, or otherwise equilibrium will not be reached at all. The critical density is given by

$$n_c \equiv \frac{n_1}{e^{\beta(\epsilon_1 - \epsilon_0)} - 1} .$$

These equations are exactly solvable for any set of parameters. For the sake of simplicity, we take $\epsilon_0 = 0$, $w = 1$ and denote $e^{-\beta(\epsilon_1 - \epsilon_0)} \equiv \xi < 1$. Then

$$n_c \equiv n_1 \frac{\xi}{1 - \xi} .$$

We eliminate f_1 in favor of f_0 through the conservation equation and get a closed equation for f_0:

$$\frac{\partial}{\partial t} f_0 = -\frac{1-\xi}{V} f_0^2 + f_0(1-\xi)(n_{tot} - n_c - \frac{1}{V(1-\xi)}) + n_{tot}. \qquad (21.15)$$

The discriminant of the polynomial (in f_0) $af_0^2 + bf_0 + c$ on the right hand side is

$$\Delta \equiv b^2 - 4ac = (1-\xi)^2(n_{tot} - n_c - \frac{1}{V(1-\xi)})^2 + 4n_{tot}\frac{1-\xi}{V} > 0,$$

with

$$b = (1-\xi)(n_{tot} - n_c - \frac{1}{V(1-\xi)}); \quad a = -\frac{1-\xi}{V} < 0; \quad c = n_{tot}.$$

The roots (x_1, x_2) are real with opposite signs

$$x_{1,2} \equiv \frac{V}{2}\left[n_{tot} - n_c - \frac{1}{V(1-\xi)} \pm \sqrt{\left(n_{tot} - n_c - \frac{1}{V(1-\xi)}\right)^2 + 4n_{tot}\frac{1-\xi}{V}} \right],$$

and the integral can be solved explicitly

$$\int dx \frac{1}{ax^2 + bx + c} = \int dx \frac{1}{a(x-x_1)(x-x_2)} = -\frac{1}{a(x_1-x_2)} \ln\left|\frac{x-x_1}{x-x_2}\right|.$$

Then, the solution (with $f_0(0) \geq 0$) is:

$$f_0(t) = \frac{x_1 + x_2 \frac{x_1 - f_0(0)}{f_0(0) - x_2} e^{-|a(x_1-x_2)|t}}{1 + \frac{x_1 - f_0(0)}{f_0(0) - x_2} e^{-|a(x_1-x_2)|t}}, \qquad (21.16)$$

and we get

$$\lim_{V\to\infty} \frac{1}{V} x_1 = \begin{cases} 0 & for & n_{tot} < n_c \\ n_{tot} - n_c & for & n_{tot} > n_c \end{cases}$$

$$\lim_{V\to\infty} \frac{1}{V} x_2 = \begin{cases} n_{tot} - n_c & for & n_{tot} < n_c \\ 0 & for & n_{tot} > n_c \end{cases}.$$

Since

$$\lim_{t\to\infty} f_0(t) = x_1,$$

we have

$$\lim_{V \to \infty} \lim_{t \to \infty} \frac{1}{V} f_0(t) = \begin{cases} 0 & for \quad n_{tot} < n_c \\ n_{tot} - n_c & for \quad n_{tot} \geq n_c \end{cases} . \tag{21.17}$$

Thus BEC occurs above the critical density.

By taking the opposite order of the limits we have to be careful regarding the initial condition. Let us consider the condensate particle density $n_0(t) \equiv \lim_{V \to \infty} \frac{1}{V} f_0(t)$.

For $n_{tot} < n_c$

$$n_0(t) = \frac{n_0(0)(n_c - n_{tot})e^{-(1-\xi)(n_c-n_{tot})t}}{n_c - n_{tot} + n_0(0)(1 - e^{-(1-\xi)(n_c-n_{tot})t})} ,$$

while for $n_{tot} > n_c$

$$n_0(t) = \frac{n_0(0)(n_{tot} - n_c)}{n_0(0) + (n_{tot} - n_c - n_0(0))e^{-(1-\xi)(n_{tot}-n_c)t}} .$$

In the special case $n_{tot} = n_c$ one has a (non-exponential) power law relaxation

$$n_0(t) = \frac{n_0(0)}{1 + n_0(0)(1 - \xi)t} ,$$

called "critical slowing down".

Therefore, if no condensate is present at the initial time ($n_0(0) = 0$), then there is no evolution of the condensate at all. For any finite initial condensate ($n_0(0) > 0$) one gets condensation at $t \to \infty$ above the critical density, while the condensate disappears below the critical density.

The modulus square of the order parameter is given just by a simple integration

$$|p(t)|^2 = |p(0)|^2 \exp \int_0^t dt' \left[b + \frac{1}{V}(1 - f_0(t')(1 - \xi)) \right] ,$$

or explicitly, using the previous solution (in the critical regime $n_{tot} > n_c$)

$$\frac{|p(t)|^2}{|p(0)|^2} = \frac{1 + \frac{x_1 - f_0(0)}{f_0(0) - x_2}}{1 + \frac{x_1 - f_0(0)}{f_0(0) - x_2}e^{-|a(x_1-x_2)|t}} e^{-(1-\xi)(\frac{x_1}{V} - n_{tot}+n_c)t} .$$

The last exponential behaves for $V \to \infty$ as $e^{-\frac{1}{V}\frac{n_c}{n_{tot}-n_c}t}$. Therefore, for any large, but finite volume, at $t \to \infty$ the order parameter disappears. However, if the infinite volume limit is performed first, then the order parameter reaches the stationary value

$$\frac{|p(\infty)|^2}{|p(0)|^2} = \frac{n_0(\infty)}{n_0(0)},$$

and the Bose-Einstein condensation is perfect also in the sense of the order parameter, if it had not been vanishing at $t = 0$. If in the initial state one had $|p(0)|^2 = n_0(0)$, then also in the final state $|p(\infty)|^2 = n_0(\infty)$. The above shown features are actually rather general and not pertinent only to this solvable model.

The figures illustrate the time evolution of $n_0(t)$ and $\frac{|p(t)|^2}{|p(0)|^2}$ above the critical density for finite volumes $V = 20, 200$. (We took here $n_{tot} = 20$; $n_c = 15$; $\frac{1}{V}f_0(0) = 0.001$; $\xi = 0.1$.)

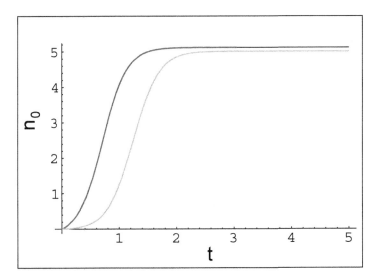

Fig. 21.2 Time evolution of the population of the lowest level above the critical density. (Lower curve for $V = 20$, upper curve for $V = 200$.)

The analysis of this solvable model shows that in a finite volume and discrete spectrum (above the critical density), there are two characteristic time scales in the evolution. A rapid volume-independent increase of the condensate in the early stage is followed by a slow decay of the condensate.

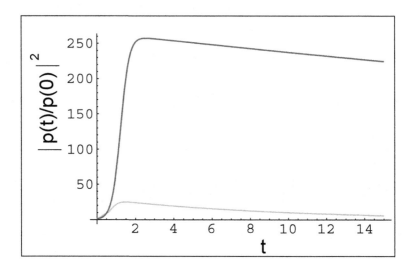

Fig. 21.3   Time evolution of the square of the order parameter. (Lower curve for $V = 20$, upper curve for $V = 200$.)

Its decay rate decreases with the volume, and for an infinitely large volume it remains hanging. This theoretical scenario may help to understand the experimentally observed BEC with spontaneous symmetry breaking (interference!) of atoms in finite magnetic traps.

*

We have seen in this Chapter that a consistent real-time description of the Bose-Einstein condensation may be formulated in the frame of Markovian dynamics for the populations and order parameter. The condensation itself is triggered in the thermodynamic limit only due to an infinitesimal initial "seed". This scenario is typical for non-linear dynamics dominated by attractors. The need for the small symmetry breaking initial condition is the analog of the infinitesimal symmetry breaking, necessary in Bogolyubov's quasi-averages formulation of the equilibrium phase transitions.

Chapter 22

Real-time experiments: Ultra-short-time spectroscopy

A new exciting sort of non-equilibrium experiments has emerged in the last decades. With the availability of lasers with carrier frequencies around the band gap of many important semiconductor materials emitting ultrashort pulses in the femtosecond range (1 fms=10^{-15} sec), we can get a "direct" glimpse on the time evolution of the state of an electron-hole plasma in a semiconductor [Schäfer, Wegener (2002)].

The simplest sort of experiment (differential transmission spectrum - DTS) consists of a strong "pump" pulse followed by a weak "test" pulse falling on a semiconductor target (see Fig. 22.1). The role of the first one is to excite electron-hole pairs, while the second one checks the state of these carriers after a certain time delay. The two pulses are either incident under a small angle, or have different polarizations, thus enabling us to define a transmitted test pulse. We can measure the spectrum of this transmitted pulse with and without the pump excitation as a function of the time delay between the pulses, and consider the existing difference. In this sense one measures indirectly the state of the system in real time.

Another type of experiment uses two incident pulses (not necessarily of different intensity) with wave vectors $\vec{k}_1$ respectively $\vec{k}_2$, but instead of measuring their transmission, one measures the spectrum of the emitted light in the direction $2\vec{k}_1 - \vec{k}_2$, which is resulting due to non-linear effects in the matter (see Fig. 22.2). Such an arrangement is called four wave mixing (FWM). One gets a so called photon-echo pulse in this new direction, delayed with twice the delay time between the pulses. His height decreases exponentially with this delay time. This experiment, since it is of interference type, contains much more interesting information about the state of the system.

The characteristic feature of this new kind of experiments is that one

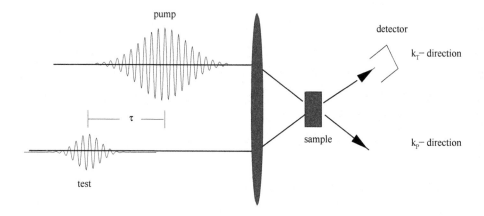

Fig. 22.1   Differential transmission configuration.

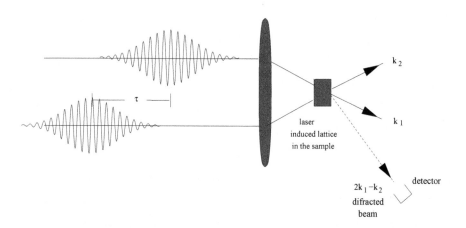

Fig. 22.2   Four-wave mixing configuration.

does not measure just a material constant, but follows the evolution of the system "in real time" from a given initial state (equilibrium) towards a new equilibrium. Thus, a theoretical description has to model the whole experiment and predict the resulting spectrum. Of course, for long time intervals (long delays between the pulses) one expects a good description in terms of Markovian processes. However, with very short pulses and very short time delays we have access also to the early coherent regime and thus to specific new effects.

A theoretical treatment of this ultra-short-time behavior needs the kind of quantum kinetic description we emphasized at the very beginning of these lectures. There are two traditional approaches to quantum kinetics. The first is based on the truncated equations of motion of different average entities we described shortly in its simplest form in Chapter 7. The second is based on the use of non-equilibrium Green-functions. In the following we want to give a glimpse at the concepts and achievements of this later approach. A thorough comparison of the two methods is difficult, but obviously the former is more appropriate in the interpretation of coherent effects, while the latter can better incorporate dissipative effects.

Chapter 23

Keldysh's non-equilibrium Green functions

The Feynman diagram technique has proved to be extremely useful in improving perturbational methods in the quantum field theory, as well as in the equilibrium theory of many-body systems. In this chapter we want to describe the special diagram technique introduced by Keldysh for the non-equilibrium Green functions. This technique follows closely the technique introduced formerly into he theory of equilibrium Green functions [Abrikosov, Gorkov, Dzyaloshinski (1965)] at zero temperature. Therefore, first we want to outline here the formalism of equilibrium Green functions in the many-body problem, already the subject of so many books in the literature.

23.1 Equilibrium Green functions

The object of main interest (in the fermionic case) is the retarded Green-function

$$G^r(\vec{x},\vec{x}';t-t') = \frac{1}{\imath}\theta(t-t')\langle\left[\psi(\vec{x},t),\psi(\vec{x}',t')^+\right]_+\rangle \,,$$

where the averaging is performed over the grand canonical ensemble

$$\langle...\rangle \equiv Tr\left\{\frac{1}{z}e^{-\beta(H-\mu N)}...\right\} \,.$$

From this one-particle Green function one can extract information about the quasiparticles or thermodynamic entities. From the two-particle Green functions (with four operators) one may compute also the linear response properties.

At $T = 0$ a Feynman diagram technique - analogous to that of the quantum field theory - exists for the causal Green function

$$G^c(\vec{x}, \vec{x}'; t - t') = \frac{1}{\imath} \langle T \left\{ \psi(\vec{x}, t), \psi(\vec{x}', t')^+ \right\} \rangle.$$

The essential ingredient is Wick's theorem, which allows a factorization of the averages of chronological products into products of propagators (one particle Green functions).

On its turn, the retarded Green function G^r in a homogeneous system, may be reconstructed from its Fourier transform in the space variable, using its relationship to the causal one

$$\tilde{G}^r(\vec{p}, \omega) = \begin{cases} \tilde{G}^c(\vec{p}, \omega) & for \quad \omega > \mu, \\ \tilde{G}^c(\vec{p}, \omega)^* & for \quad \omega < \mu. \end{cases}$$

At $T \neq 0$ one has to handle perturbatively not only the evolution operator $e^{-\imath H t}$ but also the "imaginary time" exponential $e^{-\beta H}$. Matsubara invented an analogous diagram technique, but for the imaginary-time causal Green function

$$\mathcal{G}(\vec{x}, \tau - \tau') = \frac{1}{\imath} \langle T_\tau \left\{ \psi(\vec{x}, \imath\tau) \psi(\vec{x}', \imath\tau') \right\} \rangle.$$

The essential feature one uses is the **KMS** (Kubo-Martin-Schwinger) property of these Green functions

$$\mathcal{G}(\vec{x}, \tau) = \mp \mathcal{G}(\vec{x}, \tau + \beta),$$

(the sign is negative for fermions and positive for bosons).

In this case G^r may be reconstructed from the discrete Fourier transform of the Matsubara function

$$\tilde{\mathcal{G}}(\vec{p}, \omega_n) = \tilde{G}^r(\vec{p}, \imath\omega_n); \quad (\omega_n = (2n + 1)\pi T).$$

The inconvenience is that we need the knowledge of the Matsubara function for all $n = 0, 1, 2, ..$ and one needs to suppose the good behavior of $G^r(\vec{p}, \omega)$ for $|\omega| \to \infty$ (less divergent than $e^{\pi|\omega|}$) according to Carlsons theorem.

23.2 Non-equilibrium Keldysh-Green Functions

For the treatment of a non-equilibrium problem we are interested in time-dependent non-equilibrium averages, for example:

$$\langle \psi_i(\vec{x}, t)^+ \psi_j(\vec{x}', t) \rangle = Tr\{\rho_0 \psi_i(\vec{x}, t)^+ \psi_j(\vec{x}', t)\},$$

where the indices i, j specify different particles (electrons and holes in a semiconductor,) starting from a given arbitrary initial ensemble ρ_0 (usually taken at $t = -\infty$). The information we are looking for are precisely these time dependent averages themselves. They represent - for example in an ultra-short-time spectroscopy experiment - the particle densities and the inter-band polarization, which is coupled to the light.

The idea of Keldysh was that one may construct a generalized Feynman diagram technique with analogous rules if one considers more than just the time ordered Green function G^c.

Besides the fact that the averaging is not performed on an equilibrium state, another new aspect of the non-equilibrium situation is that the time evolution occurs generally speaking with a time-dependent Hamiltonian:

$$H(t) = H_0 + H_{int}(t) = H_0 + H' + H''(t),$$

where H_0 is the Hamiltonian of the independent particles, H' is the interaction between the particles and $H''(t)$ is a time-dependent interaction with an external field. One may admit - as usual - that H' is introduced adiabatically, but $H''(t)$ is in general not adiabatic. We suppose that $H''(t) = 0$ for $t < 0$.

The unitary evolution operator $U(t)$ satisfies the Schrödinger equation

$$i\hbar \frac{\partial}{\partial t} U(t) = H(t) U(t),$$

with the usual initial condition

$$\lim_{t \to -\infty} e^{\frac{i}{\hbar} H_0 t} U(t) = 1,$$

or in the interaction picture with

$$S(t, -\infty) = e^{\frac{i}{\hbar} H_0 t} U(t),$$

$$i\hbar\frac{\partial}{\partial t}S(t,-\infty)=H_{int}(t)_{H_0}S(t,-\infty);\quad H_{int}(t)_{H_0}\equiv e^{\frac{i}{\hbar}H_0 t}H_{int}(t)e^{-\frac{i}{\hbar}H_0 t}\,,$$

and

$$S(-\infty,-\infty)=1\,.$$

The formal solution is

$$S(t,-\infty)=T\left\{e^{-\frac{i}{\hbar}\int_{-\infty}^{t}d\tau H_{int}(\tau)_{H_0}}\right\}\,.$$

One may define also

$$S(t_2,t_1)=T\left\{e^{-\frac{i}{\hbar}\int_{t_1}^{t_2}d\tau H_{int}(\tau)_{H_0}}\right\}\,,$$

with the properties $S(t_2,t_1)S(t_1,t_0)=S(t_2,t_0)$ for $t_2>t_1>t_0$.

This evolution operator connects the operators in the two pictures (supposed to be identical in the infinite past)

$$A(t)=S(t,-\infty)^{+}A_{H_0}(t)S(t-\infty)\,,$$

or in an alternative form with the help of the S-matrix $S\equiv S(-\infty,\infty)$

$$A(t)=S^{+}T\left\{A_{H_0}(t)S\right\}=\tilde{T}\left\{S^{+}A_{H_0}(t)\right\}S\,.$$

The fact that we have chosen the identity of the two pictures in the remote past has some peculiar consequences in the case of non-equilibrium, where remote past and remote future are not equivalent, as we shall see by the discussion of the time-reversal .

Let $t>t'$, then for two operators A and B

$$\begin{aligned}
A(t)B(t') &= S(t,-\infty)^{-1}A_{H_0}(t)S(t,-\infty)S(t',-\infty)^{-1}B_{H_0}(t')S(t',-\infty)\\
&= S(t,-\infty)^{-1}A_{H_0}(t)S(t,t')B_{H_0}(t')S(t',-\infty)\\
&= S(t,-\infty)^{-1}S(\infty,t)^{-1}S(\infty,t)A_{H_0}(t)S(t,t')B_{H_0}(t')S(t',-\infty)\\
&= S(\infty,-\infty)^{-1}S(\infty,t)A_{H_0}(t)S(t,t')B_{H_0}(t')S(t',-\infty)\\
&= S(\infty,-\infty)^{-1}T\left\{A_{H_0}(t)B_{H_0}(t')S(\infty,-\infty)\right\}\,.
\end{aligned}$$

Thus,

$$\langle T\left\{A(t)B(t')\right\}\rangle=\langle S^{-1}T\left\{A(t)_{:H_0}B(t')_{H_0}S\right\}\rangle\,,\tag{23.1}$$

where the inverse of the S-matrix may be expressed through the anti-chronological product

$$S^{-1} = \tilde{T}\left\{e^{\frac{i}{\hbar}\int_{-\infty}^{\infty} d\tau H_{int}(\tau)_{H_0}}\right\}.$$

The problem is that in order to obtain the Feynman diagram rules of equilibrium theory at $T = 0$ one needed the adiabaticity of the ground state $S|0\rangle = e^{i\eta}|0\rangle$ to get $\langle 0|T\{A(t)B(t')\}|0\rangle = \frac{1}{\langle 0|S|0\rangle}\langle 0|T\{A(t)_{:H_0}B(t')_{H_0}S\}|0\rangle$. This property is obvious when a gap above the ground state of the system exists (with and without interaction), since an infinitely slow perturbation cannot override this gap. For an arbitrary initial state ρ_0 (even if it would be the macro-canonical equilibrium of the unperturbed system) and a time dependent non-adiabatic perturbation, however, an analogous statement does not hold.

Nevertheless, Keldysh assumes in his original paper that until the introduction of the time dependent (non-adiabatical) perturbation at some positive time a general adiabatic evolution of the equilibrium state

$$S(0,-\infty)\frac{e^{-\beta(H_0-\mu_0 N)}}{Z_0}S(0,-\infty)^{-1} = \frac{e^{-\beta(H-\mu N)}}{Z},$$

still holds. Such a statement, while not necessary for the diagram technique, would establish the connection to the equilibrium Green functions

$$Tr\left\{\frac{e^{-\beta(H-\mu N)}}{Z}T\{\psi(\vec{x},t),\psi(\vec{x}',t')^+\}\right\} \doteq$$

$$Tr\left\{\frac{e^{-\beta(H_0-\mu_0 N)}}{Z_0}T_\mathcal{C}\{\psi(\vec{x},t),\psi(\vec{x}',t')^+ S_\mathcal{C}\}\right\}.$$

However, the validity of such an assumption is highly questionable [Rammer and Smith (1986)].

Here we are not interested as much in calculating with the new technique equilibrium Green functions at non-zero temperature as in true non-equilibrium situations. For this purpose we write down Eq.(23.1) in a different form. Namely, as a "chronological" product on a Keldysh-contour $\mathcal{C}$ made of two branches as shown in Fig. (23.1). The first branch runs from $-\infty$ to ∞, where the physical times t, and t' lie, and a second branch

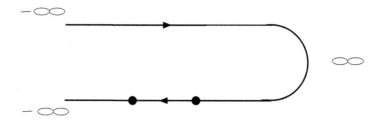

Fig. 23.1   Keldysh contour

running from ∞ to $-\infty$. Then

$$\langle T\left\{A(t)B(t')\right\}\rangle = \langle T_C\left\{A(t)_{:H_0}B(t')_{H_0}S_C\right\}\rangle\,,$$

with

$$S_C = T_C\left\{e^{-\frac{i}{\hbar}\int_C d\tau H_{int}(\tau)_{H_0}}\right\}\,,$$

where $\int_C d\tau \equiv \sum_{\eta=\pm 1}\eta\int_{-\infty}^{\infty}d\tau$ and on the upper branch ($\eta = -1$) one has anti-chronological ordering. It is worth noticing that the averaging is made always over the state ρ_0. Obviously

$$\langle S_C\rangle = 1\,,$$

and Wick's theorem is valid if one considers a matrix of Green functions corresponding to the different situations of the times on the contour

$$G_C(t,t') = \begin{pmatrix} G^c(t,t') & G^<(t,t') \\ G^>(t,t') & G^{\tilde{c}}(t,t') \end{pmatrix}\,;$$

with

$$G^<(t,t') = i\langle\psi(t')^+\psi(t)\rangle\,,$$
$$G^>(t,t') = -i\langle\psi(t)\psi(t')^+\rangle\,,$$

and

$$G^c(t,t') = \theta(t-t')G^>(t,t') + \theta(t'-t)G^<(t,t')\,,$$
$$G^{\tilde{c}}(t,t') = \theta(t-t')G^<(t,t') + \theta(t'-t)G^>(t,t')\,.$$

This means that the average of the new chronological product introduced by Keldysh may be decomposed in a product of these new propagators.

Remark that the new propagators (one particle Green functions) now depend on both time-variables.

Actually, it is more useful to introduce only three independent Green functions , whereas all the others may be obtained by their linear combination. They are the retarded and advanced Green functions

$$G^r(t, t') = \theta(t - t') \left(G^>(t, t') - G^<(t, t') \right) ,$$

$$G^a(t, t') = \theta(t' - t) \left(G^<(t, t') - G^>(t, t') \right) ,$$

together with

$$G^\triangle = -G^>(t, t') - G^<(t, t') .$$

We'd like to mention that higher Green functions may be also introduced, as in the equilibrium theory.

Chapter 24

Keldysh-Green functions and time-reversal

In order to understand the problem of the approach to equilibrium within the various approximation schemes, it is important to know the time-reversal property of the non-equilibrium Green functions. Since in interacting many-body systems the second quantization formalism is an essential ingredient, we discuss time-reversal invariance within this formalism. The treatment is borrowed from the quantum field theory [Lee (1981)], by omitting some aspects related to the relativistic formulation.

In ordinary quantum mechanics of the motion of a particle in external electric and magnetic fields, the basic statement about the time reversal stems from the invariance of the Hamiltonian

$$\mathcal{H} = \frac{1}{2m}\left(\imath\hbar\nabla - \frac{e}{c}\vec{A}\right)^2 + eV$$

with respect to a complex conjugation and a simultaneous reversal of the magnetic fields

$$\mathcal{H}^*_{-\vec{A}} = \mathcal{H}_{\vec{A}}\,.$$

As a consequence, if $\psi(t)$ is a solution of the Schrödinger equation

$$\imath\hbar\frac{\partial}{\partial t}\psi = \mathcal{H}\psi\,,$$

the wave function

$$\psi' \equiv \psi^*(-t)_{-\vec{A}}$$

187

is also a solution. Obviously, by this operation of complex conjugation and magnetic field reversal, the momentum and angular momentum operators - like their classical counterparts - change their signs.

In the formulation of time reversal within the second quantization formalism, we start from this last property. We consider as a working example the Hamiltonian of interacting charged particles and photons Eq.(16.4) together with the terms representing the interaction with some external electromagnetic sources Eq.(16.4). The time dependence of this Hamiltonian comes here only through the time dependence of the external sources.

Now, let us define a unitary operator U_T which changes the signs of the momentum and angular momenta of the particles as

$$U_T a_{\vec{k},\sigma_z} U_T^+ = a_{-\vec{k},-\sigma_z}; \quad U_T b_{\vec{q},s} U_T^+ = -b_{-\vec{q},-s} \, .$$

Here the index s of the photonic operators represent the helicity ($s = \pm$), and we omitted the index α specifying the different sorts of particles. Then,

$$U_T \psi_\sigma(\vec{r}) U_T^+ = \psi_{-\sigma}(\vec{r})^*; \quad U_T \vec{A}(\vec{r}) U_T^+ = -\vec{A}(\vec{r})^* \, ,$$

where the star means complex conjugation of all the c-numbers (here the plane wave and polarization-vector factors) without touching the creation and annihilation operators. As a consequence, the action of the unitary transformation U_T on the Hamiltonian of the fermions and photons Eq.(16.3) is equivalent again to a complex conjugation

$$U_T H_{fp} U_T^+ = H_{fp}^* \, ,$$

while in order to obtain the same effect in the external part Eq. (16.4) we need to reverse also the sign of the external current. Thus, for the total Hamiltonian Eq.(16.5) we have the time-reversal invariance as

$$\left. \left(U_T H(t) U_T^+ \right)^* \right|_{-j_{ext}} = \left. H(t) \right|_{j_{ext}} \, . \tag{24.1}$$

If instead of the unitary operator U_T one introduces an anti-unitary (unitary but anti-linear) operator T such that its action incorporate both the unitary transformation and the complex conjugation, the time-reversal invariance may be formulated as

$$\left. T H(t) T^{-1} \right|_{-j_{ext}} = \left. H(t) \right|_{j_{ext}} \, . \tag{24.2}$$

Let us admit that the unperturbed part H_0 of the Hamiltonian does not include external currents and therefore is time-reversal invariant, i.e.

$$U_T H_0^* U_T^+ = H_0 .$$

In the interaction picture the perturbation is

$$H_{int}(t)_{H_0} = e^{\frac{\imath H_0 t}{\hbar}} H_{int}(t) e^{-\frac{\imath H_0 t}{\hbar}} ,$$

and the time-reversal property looks as

$$U_T \left. H_{int}(t)_{H_0}^* \right|_{j_{ext}, \rho_{ext}} U_T^+ = \left. H_{int}(-t)_{H_0} \right|_{j_{ext}^T, \rho_{ext}} , \qquad (24.3)$$

with

$$\vec{j}_{ext}(t)^T \equiv -\vec{j}_{ext}(-t); \quad \rho_{ext}(t)^T = \rho_{ext}(-t) .$$

This peculiar transformation is due to the fact that one has to deal with two different time-dependences: one introduced by the transformation to the interaction picture and another, due to the explicit time-dependence of the external current.

From this property Eq. (24.3) it follows the transformation for the S-matrix:

$$U_T \left. S^* \right|_{j_{ext}} U_T^+ = T \left\{ e^{\frac{\imath}{\hbar} \int_{-\infty}^{\infty} d\tau \, H_{int}(-\tau)_{H_0} \big|_{j_{ext}^T, \rho_{ext}^T}} \right\}$$

$$= \tilde{T} \left\{ e^{\frac{\imath}{\hbar} \int_{-\infty}^{\infty} d\tau \, H_{int}(\tau)_{H_0} \big|_{j_{ext}^T, \rho_{ext}^T}} \right\}$$

$$= \left. S^+ \right|_{j_{ext}^T, \rho_{ext}^T} .$$

We can now apply this relationship to derive the time-reversal property of the Keldysh-Green functions. Of course, the time-reversal property of a time-dependent average depends not only on the properties of the Hamiltonian, but also on that of the state, on which the averaging occurs. We admit that this state - described by the density operator ρ_0 - is T - invariant (without being necessarily an equilibrium state), i.e.

$$U_T \rho_0^* U_T^+ = \rho_0 . \qquad (24.4)$$

In order to simplify the problem, we shall consider that our Hamiltonian is spin-conserving and invariant to spin reflection, and we suppose that the initial state is also invariant to spin reflection. In this case we have

$G_{\sigma\sigma'}(\vec{r},t;\vec{r}',t') = \delta_{\sigma,\sigma'}G(\vec{r},t;\vec{r}',t')$ and we may drop in what follows the spin index.

Now, let us consider for example the Green function

$$G^>(\vec{r},t;\vec{r}',t') = -\imath\langle\psi(\vec{r},t)\psi^+(\vec{r}',t')^+\rangle\,.$$

Using the time reversal property of the initial state Eq.(24.4) one gets for any operator A

$$\langle A\rangle = \langle U_T A^* U_T^+\rangle^*\,,$$

or using also the hermiticity of ρ_0

$$\langle A\rangle = \langle U_T A^{+*} U_T^+\rangle\,.$$

Now, with our definition of the connection between the Heisenberg and interaction pictures we have

$$G^>(\vec{r},t;\vec{r}',t') = -\imath\langle\left(\tilde{T}\left\{S^+\psi_{H_0}(\vec{r},t)\right\}T\left\{\psi_{H_0}(\vec{r}',t')^+S\right\}\right)\rangle\,,$$

and therefore, using the above property of the averages and the action of time reversal on the operators

$$G^>(\vec{r},t;\vec{r}',t') = -\imath\left\langle\left(T\left\{\psi_{H_0}(\vec{r},-t)S\right\}\tilde{T}\left\{\psi_{H_0}(\vec{r}',-t')^+S^+\right\}\right)^+\right\rangle\Bigg|_{j_{ext}^T,\rho_{ext}^T}$$

$$= -\imath\left\langle T\left\{\psi_{H_0}(\vec{r}',-t')S\right\}\tilde{T}\left\{\psi_{H_0}(\vec{r},-t)^+S^+\right\}\right\rangle\Bigg|_{j_{ext}^T,\rho_{ext}^T}$$

$$= -\imath\left\langle S\psi(\vec{r}',-t')\psi(\vec{r},-t)^+S^+\right\rangle\Big|_{j_{ext}^T,\rho_{ext}^T}\,.$$

The S-matrices in the last line may be interpreted as performing the average on a new density operator

$$\tilde{\rho}_0 = S^+\rho_0 S\,,$$

instead of ρ_0 . This density operator describes the state of the system in the remote future. The existence of such a state - as well as that of the S-matrix itself - is not obvious, and implies necessarily the decoupling of the time dependent external sources after a finite time. This situation occurs in the case of short optical pulses. In the absence of the time-dependent sources, at $T = 0$, due to the adiabatic theorem, the final state coincides with the initial one, and the effect of time reversal is more simple.

The result may be written generally for the whole matrix of Keldysh-Green functions as

$$G(\vec{r}, t; \vec{r}', t')|_{j_{ext}, \rho_{ext}, \rho_0} = G(\vec{r}', -t'; \vec{r}, -t)|_{j_{ext}^T, \rho_{ext}^T, \tilde{\rho}_0} . \tag{24.5}$$

This is the time-reversal covariance of the Keldysh-Green functions. However, under this form, it is not yet very useful, since we still need a recipe to calculate the averages on $\tilde{\rho}$. In order to get this recipe, let us define another similar set of Green functions, this time on the Keldysh anti-contour $\tilde{C}$ as shown in Fig. (24.1). According to the time-ordering on this new contour we have

$$G_{\tilde{C}}(t, t') = \begin{pmatrix} G^{\tilde{c}}(t, t') & G^>(t, t') \\ G^<(t, t') & G^c(t, t') \end{pmatrix} .$$

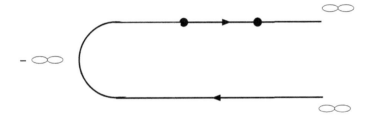

Fig. 24.1   Keldysh anti-contour

This is actually equivalent to considering the old contour, but the average taken on the final density operator $\tilde{\rho}_0$ instead of the initial one ρ_0. In terms of the G^r, G^a and $G^\triangle$ Green functions the relationship reads as

$$G_{\tilde{C}}^{r,a}\Big|_{\rho_0} = G_C^{a,r}\Big|_{\rho_0} ; \quad G_{\tilde{C}}^\triangle\Big|_{\rho_0} = G_C^\triangle\Big|_{\rho_0} .$$

In the non-interacting case one has just

$$G_{\tilde{C}}^{0r,a} = G_C^{0a,r}; \quad G_{\tilde{C}}^{0\triangle} = G_C^{0\triangle} .$$

These last relationships mean that the diagrammatic rules with ρ_0 on the new contour $\tilde{C}$ may be obtained from the old rules with ρ_0 on the old contour C by just exchanging G_C^{0a} with G_C^{0r}. On the other hand, the total Green functions with $\tilde{\rho}_0$ on the old contour C may be obtained also by the same permutation of the retarded-advanced indices from those with ρ_0 on the new contour $\tilde{C}$, by the same permutation of indices. Therefore, if the original Green functions are known as a given functional of the free propagators

$$G_C^{r,a,\triangle}\Big|_{\rho_0} = \mathcal{G}^{r,a,\triangle}\left[G_C^{0r}, G_C^{0a}, G_C^{0\triangle}\right],$$

then, by a double change of retarded-advanced indices, one gets

$$G_C^{r,a,\triangle}\Big|_{\tilde{\rho}_0} = \mathcal{G}^{a,r,\triangle}\left[G_C^{0a}, G_C^{0r}, G_C^{0\triangle}\right].$$

Now we may reformulate the basic time-covariance of the Green functions Eq.(24.5) as

$$\mathcal{G}^{r,a,\triangle}\left(\vec{r}, t : \vec{r}', t'; \left[G_C^{0r}, G_C^{0a}, G_C^{0\triangle}, j_{ext}, \rho_{ext}\right]\right)$$

$$= \mathcal{G}^{r,a,\triangle}\left(\vec{r}', -t' : \vec{r}, -t; \left[G_C^{0a}, G_C^{0r}, G_C^{0\triangle}, j_{ext}^T, \rho_{ext}^t\right]\right).$$

This is an explicit functional relationship obeyed by the Keldysh-Green functions in virtue of the time-reversal covariance Eq.(24.1) of the Hamiltonian.

Chapter 25

Time-dependent Hartree-Fock treatment of Coulomb interaction. The exciton

Within the described diagram technique, the starting point for the calculation of the two-point Green functions is the Dyson equation

$$G(t,t') = G^{(0)}(t,t') + \int dt'' \int dt''' G^{(0)}(t,t'')\Sigma(t'',t''')G(t''',t') \,,$$

where the integration occurs over the time-contour $\mathcal{C}$, and $G^{(0)}(t,t')$ is the free Green function. Obviously, we have to write down the equation for the time variables lying on each possible contour branch. The choice of the self-energy Σ defines the given approximation. With a simplified symbolic notation we have

$$G = G^{(0)} + G^{(0)}\Sigma G = G^{(0)} + G\Sigma G^{(0)} \,.$$

Fig. 25.1 Dyson equation for the Green functions.

If we apply the general theory to electrons in the conduction band and holes in the valence band, the Green functions and the self-energy carry also band indices $\mu, \nu = c, v$. The differential form of the Dyson equation for the Green function of wave vector $\vec{k}$ in a homogeneous system is

$$\left(\imath\hbar\frac{\partial}{\partial t} - \epsilon_{\vec{k},\mu}\right)G_{\vec{k},\mu\nu}(t,t') = \delta_c(t-t')\delta_{\mu\nu} + \Sigma_{\vec{k},\mu\lambda}(t,t_1)G_{\vec{k},\lambda\nu}(t_1,t'),$$

$$\tag{25.1}$$

$$\left(-\imath\hbar\frac{\partial}{\partial t'} - \epsilon_{\vec{k},\nu}\right)G_{\vec{k},\mu\nu}(t,t') = \delta_c(t-t')\delta_{\mu\nu} + G_{\vec{k},\mu\lambda}(t,t_1)\Sigma_{\vec{k},\lambda\nu}(t_1,t').$$

Summation over the internal band index λ and integration of internal time t_1 are assumed. The contour δ function $\delta_c(t-t')$ is defined as $\pm\delta(t-t')$ for t,t' both on the upper/lower contour branch and zero if the time arguments are on different branches.

We always start from the e-h vacuum at $t = -\infty$, and at a finite time introduce an optical pulse which creates e-h pairs, which at their turn interact through Coulomb forces and with phonons. In this case the free Green function is diagonal in the band index, and its Fourier transform in the space variable is

$$G_C^{(0)}(t,t')_{\vec{p},cc} = \frac{1}{\imath}e^{-\frac{\imath}{\hbar}\epsilon_{c,\vec{p}}(t-t')}\begin{pmatrix}\theta(t-t') & 0 \\ 1 & \theta(t'-t)\end{pmatrix},$$

$$G_C^{(0)}(t,t')_{\vec{p},vv} = \frac{1}{\imath}e^{-\frac{\imath}{\hbar}\epsilon_{v,\vec{p}}(t-t')}\begin{pmatrix}-\theta(t'-t) & -1 \\ 0 & -\theta(t-t')\end{pmatrix},$$

with parabolic band spectra $\epsilon_{c,\vec{p}} = \frac{1}{2}E_g + \frac{\hbar^2}{2m_e}\vec{p}^2$ and $\epsilon_{v,\vec{p}} = -\frac{1}{2}E_g - \frac{\hbar^2}{2m_h}\vec{p}^2$.

We have to take into account that the electron-hole pairs are created by an optical pulse having the carrier frequency ω and the envelope $\mathcal{E}(t)$ through an inter-band dipole-coupling in the rotating wave approximation

$$-d\mathcal{E}(t)e^{\imath\omega t}\sum_{\vec{p}} a_{c\vec{p}}^+ a_{v\vec{p}} + h.c.,$$

corresponding to the external-field self-energy diagram of Fig.(25.2). Here - as it is usual in the treatment of inter-band absorption and emission in a semiconductor - we have neglected the photon momentum, since it is extremely small. We can see within this approximation that the one-particle part of the Hamiltonian is equivalent to a sum of two-level systems, the same we analyzed earlier in Chapter 18. The new aspect is that the many-body interactions couple these two-level systems.

Fig. 25.2   External field self-energy.

The simplest way to take into account Coulomb interactions is to consider for the self-energy the simple Keldysh-Feynman diagrams with full particle propagators and bare Coulomb line (see Fig.25.3) corresponding to the time-dependent Hartree-Fock Hamiltonian.

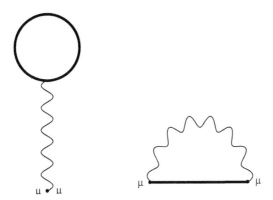

Fig. 25.3   Hartree and Fock self-energies.

The first diagram (Hartree term) vanishes due to charge neutrality of the homogeneous system. Since the bare Coulomb propagator is instantaneous in time, no time integrations survive in the Dyson equation. It is convenient to consider the differential form of the Dyson equation by taking its time derivative. Since we are mainly interested in the averages $\langle a^{+}_{\mu\vec{p}}(t)a_{\nu\vec{p}}(t)\rangle = \frac{1}{i}G^{<}_{\vec{p}}(t,t)_{\mu\nu}$, we are going to write down directly the time evolution equations for these entities, by taking into account that $\frac{\partial}{\partial t}G(t,t) = \frac{\partial}{\partial t}\left(G(t,t') + G(t',t)\right)_{t'=t}$. The inhomogeneous delta function terms disappear from this combination, and we get

$$\left(i\hbar \frac{\partial}{\partial t} - \epsilon_{\mu,\vec{p}} + \epsilon_{\nu,\vec{p}} \right) G_{\vec{p}}^{<}(t,t)_{\mu\nu}$$

$$= -d\mathcal{E}(t)e^{i\omega t}(1 - \delta_{\mu\nu}) + \frac{i}{\Omega} \sum_{\vec{q}} V_{\vec{q}} \left[G_{\vec{p}-\vec{q}}^{<}(t,t)_{\mu\nu} + \frac{1}{i}\delta_{\mu v}\delta_{\nu v} \right].$$

(The very last term is due to the compensating positive background charge.) As here we work with an effective one-particle Hamiltonian we get a closed system of equations. The Fock terms lend these equations a peculiar feature. The whole Coulomb spectrum, including the bound states (exciton), is present in this equation, although the HF-Hamiltonian has only renormalized independent momentum state particles. This feature is known also in the frame of the linear response within the HF approximation.

Chapter 26

Quantum kinetics of electrons and holes scattered by LO phonons within the s.c. RPA and the Kadanoff-Baym approximation

Let us consider in this chapter a quantum kinetic approach to the evolution of the electrons and holes created by an ultra-short optical pulse. Here we take into account only the interaction of the carriers with LO-phonons, and neglect the Coulomb interaction. The treatment will be essentially simplified by reducing the true two-time problem of the Green functions to equations for one-time averages through the Kadanoff-Baym approximation. These simplifications are justified for low electron-hole densities and weak polar couplings.

The Fröhlich interaction of electrons (or holes) with LO-phonons in a polar semiconductor is given by the interaction Hamiltonian

$$H_{int} = \sum_{q,k} g_q a_k^+ a_{k-q} (b_q + b_{-q}^+) \, .$$

Fig. 26.1   LO-phonon RPA self-energy.

The RPA self-energy diagram of Fig.(26.1) leads to the supplementary collision term

$$\sum_{\rho} \int_{-\infty}^{t} dt' \Big[\Sigma^{>}_{\mu\rho,\vec{k}}(t,t')_{phon} G^{<}_{\rho\nu,\vec{k}}(t',t) - \Sigma^{<}_{\mu\rho,\vec{k}}(t,t')_{phon} G^{>}_{\rho\nu,\vec{k}}(t',t)$$

$$- G^{>}_{\mu\rho,\vec{k}}(t,t') \Sigma^{<}_{\rho\nu,\vec{k}}(t',t)_{phon} + G^{<}_{\mu\rho,\vec{k}}(t,t') \Sigma^{>}_{\rho\nu,\vec{k}}(t',t)_{phon} \Big] ,$$

with the self-energy

$$\Sigma^{\lessgtr}_{\mu\rho,\vec{k}}(t_1,t_2)_{phon} = \frac{\imath}{\Omega} \sum_{\vec{q}} g_{\vec{q}}^2 \, D^{\lessgtr}_{\vec{q}}(t_1,t_2) \, G^{\lessgtr}_{\mu\rho,\vec{k}-\vec{q}}(t_1,t_2) .$$

We consider here only free phonon-propagators at a finite temperature T

$$D^{\lessgtr}_{\vec{q}}(t_1,t_2) = -i \sum_{\pm} N^{\pm}_{\vec{q}} e^{\pm i\omega_0(t_1-t_2)} ,$$

with

$$N^{\pm}_{\vec{q}} = N + \frac{1}{2} \pm \frac{1}{2} ; \qquad N = 1/(\exp(\omega_0/k_{\rm B}T) - 1) .$$

Unfortunately, the "collision term" for the time-diagonal Green function contains Green functions that are off-diagonal in time. We are going to make further approximations, in order to close the equation for time-diagonal Green functions. In equilibrium, a simple relationship between different Green functions takes place. It is assumed that the same relationship exist also in the weakly interacting system in non-equilibrium. This Generalized Kadanoff-Baym-Ansatz looks as:

$$G^{\lessgtr}_{\mu\nu,\vec{k}}(t_1,t_2) \simeq i \sum_{\sigma} \Big\{ G^{r}_{\mu\sigma,\vec{k}}(t_1,t_2) G^{\lessgtr}_{\sigma\nu,\vec{k}}(t_2,t_2) - G^{\lessgtr}_{\mu\sigma,\vec{k}}(t_1,t_1) G^{a}_{\sigma\nu,\vec{k}}(t_1,t_2) \Big\} .$$

For the retarded Green function

$$G^{r}_{\mu\nu,\vec{k}}(t_1,t_2) = -i\theta(t_1 - t_2)\langle [a_{\mu,\vec{k}}(t_1), a^{\dagger}_{\nu,\vec{k}}(t_2)]_+ \rangle$$

one takes usually the simplest approximation

$$G^{r}_{\mu\nu,\vec{k}}(t_1,t_2) \approx -i\delta_{\mu\nu}\theta(t_1 - t_2) e^{\frac{i}{\hbar}(\epsilon_{\vec{k}}+i\gamma)t} ,$$

where a life-time factor is incorporated.

Within a more sophisticated approximation one may also take into account in the retarded Green functions the whole evolution, due to the effective one-particle part of the Hamiltonian described in the previous chapter.

Under these conditions we get a closed system of (integro-differential) equations for the electron-hole populations and the inter-band polarization, defined for convenience as elements of the reduced density matrix $\rho_{\mu\nu,\vec{k}}(t)$

$$G^<_{\mu\nu,\vec{k}}(t,t) = i\rho_{\mu\nu,\vec{k}}(t) = i\langle a^\dagger_{\nu,\vec{k}}(t) a_{\mu,\vec{k}}(t)\rangle\,,$$

$$G^>_{\mu\nu,\vec{k}}(t,t) = -i\left(\delta_{\mu\nu} - \rho_{\mu\nu,\vec{k}}(t)\right)\,. \tag{26.1}$$

Then we get a generalization of the Bloch equations for a two-level atom of Chapter 18, this time with the many-body collision terms

$$-\frac{\partial \rho_{\mu\nu,k}}{\partial t}\bigg|_{scatt} = -\sum_{q,\sigma\tau,\pm}\Bigg\{g_q^2 \int_{-\infty}^{t} dt' \Big\{ G^r_{\mu\sigma,k-q}(t,t') G^a_{\tau\nu,k}(t',t) e^{\pm i\omega_{LO}(t-t')}$$

$$\times \Big[N_q^\mp \rho_{\sigma\tau,k}(t') - N_q^\pm \rho_{\sigma\tau,k-q}(t') \pm \sum_\rho \rho_{\sigma\rho,k-q}(t')\rho_{\rho\tau,k}(t') \Big]\Big\}$$

$$-\left\{\vec{k}\leftrightarrow\vec{k}-\vec{q}\right\}\Bigg\}\,. \tag{26.2}$$

In order to obtain experimental predictions, the so derived equations must have to be considered - as in the case of atoms - together with the Maxwell equations for the electromagnetic field. We are not going to analyze here those aspects, we concentrate only on the quantum kinetics of the carriers.

Unfortunately, we know almost nothing about the general properties of the solutions, except the conservation of the average number of particles after the optical pulse. If however, $\frac{\hbar}{\gamma}$ is chosen less or equal to the golden-rule life-time of the state $\vec{k}$, then in all numerical studies $0 \le f_e, f_h \le 1$ as it should be.

The only justification of all the approximations may be the phenomenological success of such calculations, as only numerical results are available. As we have already mentioned, such calculations include also an electromagnetic analysis of the emitted and absorbed waves. The good agreement of the quantum kinetic predictions with the experimental curves [Bányai, Tran Thoai, Reitsamer, Haug, Steinbach, Wehner, Wegener, Marschner and Stolz (1995)] is remarkable, as we illustrate here on the example of the

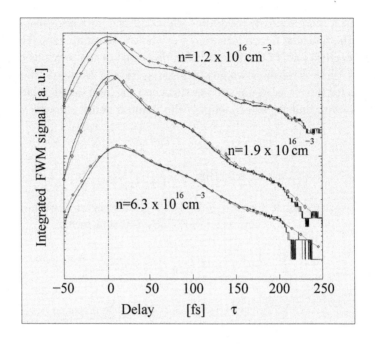

Fig. 26.2   FWM signal (experiment and theory).

prediction for the Four-Wave-Mixing (see Fig. 26.2) at low carrier electron-hole densities in GaAs, a direct gap semiconductor with a weak coupling to the LO-phonons (for electrons $\alpha \approx 0.06$). The slowly oscillating decay is a typical feature, which cannot be explained without the non-Markovian memory effects incorporated in the quantum kinetics.

Chapter 27

Full two-time quantum-kinetics of electrons and holes scattered by LO phonons within the s.c. RPA

We have seen in the previous Chapter that for a weak coupling to the LO-phonons, the reduction of the true two-time quantum kinetics of Green functions to averages depending on a single time through the Kadanoff-Baym-Ansatz, leads to good phenomenological predictions. However, one could expect that for stronger electron-phonon coupling this approximation does not work. Therefore, in this chapter we shall renounce to this approximation, and formulate a true two-time quantum kinetics. We analyze also the delicate problem of the initial conditions in this non-trivial mathematical problem with two time variables [Gartner, Bányai, Haug (1999)].

We begin with reducing the number of independent variables by using their symmetry relations

$$G_{\vec{k}}^{\gtrless}(t,t')^*_{\mu\nu} = -G_{\vec{k}}^{\gtrless}(t',t)_{\mu\nu} \,,$$

$$G_{\vec{k}}^{R}(t,t')^*_{\mu\nu} = G_{\vec{k}}^{A}(t',t)_{\mu\nu} \,.$$

Similar relations hold for the self-energies Σ too. From now on, one may restrict itself to $t \geqslant t'$. In this case it is natural to work instead of t' with $\tau \equiv t - t' \geq 0$. With the help of the symmetry relations we can bring the collision terms to a form where only Green functions appear with the first time argument greater than the second

$$i\hbar\frac{\partial}{\partial t}G_{\vec{k}}^{r}(t,t-\tau)\Big|_{coll} = \int_{t-\tau}^{t}dt'\left[\Sigma_{\vec{k}}^{r}(t,t')G_{\vec{k}}^{r}(t',t-\tau) - G_{\vec{k}}^{r}(t,t')\Sigma_{\vec{k}}^{r}(t',t-\tau)\right],$$

$$(27.1)$$

and

$$\imath\hbar\frac{\partial}{\partial t}G_{\vec{k}}^{<}(t, t-\tau)\bigg|_{coll} \tag{27.2}$$

$$= \int_{-\infty}^{t-\tau} dt' \left[-\Sigma_{\vec{k}}^{r}(t, t')G_{\vec{k}}^{<}(t-\tau, t')^{+} + \Sigma_{\vec{k}}^{<}(t, t')G_{\vec{k}}^{r}(t-\tau, t')^{+} \right.$$

$$\left. + G_{\vec{k}}^{r}(t, t')\Sigma_{\vec{k}}^{<}(t-\tau, t')^{+} - G_{\vec{k}}^{<}(t, t')\Sigma_{\vec{k}}^{r}(t-\tau, t')^{+} \right]$$

$$+ \int_{t-\tau}^{t} dt' \left[\Sigma_{\vec{k}}^{r}(t, t')G_{\vec{k}}^{<}(t', t-\tau) - G_{\vec{k}}^{r}(t, t')\Sigma_{\vec{k}}^{<}(t', t-\tau) \right] .$$

The self-energies are given by

$$\Sigma_{\vec{k}}^{r}(t, t-\tau) = \frac{\imath\hbar}{\Omega} \sum_{\vec{q}} g_{\vec{q}}^{2} \left[D_{\vec{q}}^{>}(\tau)G_{\vec{k}-\vec{q}}^{r}(t, t-\tau) + D_{\vec{q}}^{r}(\tau)G_{\vec{k}-\vec{q}}^{<}(t, t-\tau) \right] ,$$

$$\Sigma_{\vec{k}}^{<}(t, t-\tau) = \frac{\imath\hbar}{\Omega} \sum_{\vec{q}} g_{\vec{q}}^{2} \ D_{\vec{q}}^{<}(\tau)G_{\vec{k}-\vec{q}}^{<}(t, t-\tau) . \tag{27.3}$$

The phonon propagators, as before, will be taken to be the equilibrium ones, i.e. the phonons are treated as a thermal bath not influenced by the electrons and holes

$$D_{\vec{q}}^{\gtrless}(\tau) = \frac{1}{\imath\hbar} \sum_{\xi=\pm 1} N_{\xi} e^{\mp\imath\xi\omega_0\tau} ,$$

$$D_{\vec{q}}^{r}(\tau) = \frac{1}{\imath\hbar} \sum_{\xi=\pm 1} \xi e^{-\imath\xi\omega_0\tau} , \tag{27.4}$$

where $N_{-1} = N$ is the Bose function and $N_{+1} = N + 1$.

As we can see, the collision terms contain only Green functions where both time arguments refer to earlier times as the ones at the left hand side.

Let us assume now that the optical pulse begins at $t = t_0$. Until this time the system is supposed to have been in equilibrium with no electrons or holes, and the phonons being at a temperature T. There is some discrepancy in this assumption, due to the fact that at any finite temperature some carriers are always present, but their density is so small that we may neglect it. Under this condition, at any $t < t_0$ (with $\tau > 0$) we have equilibrium Green functions. Since we do not want to restrict ourselves here to very weak phonon coupling, we have to take into account the phonon interaction also for this specific equilibrium problem; this is a typical polaronic effect. With these assumptions, the only surviving Green function for

$t' < t < t_0$ is the retarded one (polaron Green function), coinciding with the causal one:

$$G_{\vec{k}}^r(t, t - \tau) \equiv \mathcal{G}_{\vec{k}}^r(\tau) = \begin{pmatrix} \mathcal{G}_{\vec{k}c}^r(\tau) & 0 \\ 0 & \mathcal{G}_{\vec{k}v}^r(\tau) \end{pmatrix} ;$$

$$G_{\vec{k}}^<(t, t - \tau) \equiv \mathcal{G}_{\vec{k}}^<(\tau) = \begin{pmatrix} 0 & 0 \\ 0 & -\mathcal{G}_{\vec{k}v}^r(\tau) \end{pmatrix} ;$$

$$\Sigma_{\vec{k}}^r(t, t - \tau) = \sigma_{\vec{k}}^r(\tau) = \begin{pmatrix} \sigma_{\vec{k}c}^r(\tau) & 0 \\ 0 & \sigma_{\vec{k}v}^r(\tau) \end{pmatrix} ;$$

$$\Sigma_{\vec{k}}^<(t, t - \tau) = \sigma_{\vec{k}}^<(\tau) = \begin{pmatrix} 0 & 0 \\ 0 & -\sigma_{\vec{k}v}^r(\tau) \end{pmatrix} ,$$

having the properties

$$\mathcal{G}_{\vec{k}v}^r(\tau) = -\mathcal{G}_{\vec{k}h}^r(\tau)^* ; \quad \sigma_{\vec{k}v}^r(\tau) = -\sigma_{\vec{k}h}^r(\tau)^* .$$

Denoting (for electrons or holes):

$$\mathcal{G}_{\vec{k}}^r(\tau) = \frac{1}{\imath} \mathcal{G}_{\vec{k}}(\tau)\theta(\tau) e^{-\frac{\imath}{\hbar}\epsilon_{\vec{k}}\tau} ,$$

one has:

$$\frac{\partial}{\partial \tau}\mathcal{G}_{\vec{k}}(\tau) = \frac{1}{\imath\hbar} \int_0^t d\tau' e^{\frac{\imath}{\hbar}\epsilon_{\vec{k}}(\tau - \tau')} \sigma_{\vec{k}}^R(\tau - \tau')\mathcal{G}_{\vec{k}}(\tau') , \qquad (27.5)$$

with the kernel

$$\sigma_{\vec{k}}^R(\tau) = \frac{1}{\imath\Omega} \sum_{\vec{q},\xi=\pm} g_{\vec{q}}^2 N_\xi \mathcal{G}_{\vec{k}-\vec{q}}(\tau) e^{-\frac{\imath}{\hbar}(\epsilon_{\vec{k}-\vec{q}} + \xi\hbar\omega_0)\tau} , \qquad (27.6)$$

and the boundary condition

$$\mathcal{G}_{\vec{k}}(0) = 1 \,. \tag{27.7}$$

Thus, the initial condition for the two-time problem is defined also by a non-linear integro-differential equation.

The rapid technical progresses in the computer speed and memory allowed also numerical solutions of the equations. The numerical strategy to solve the equations is illustrated in Fig.27.1 of the two times plane. After a discretization in time, one first solves numerically the one-time problem defined by Eqs.(27.5) -(27.7). This means the knowledge of the Green functions inside the lower triangle below the point (t_0, t_0), going by vertical steps downstairs. Thereafter one makes diagonal steps starting for every time t according to the structure of the Eqs.(27.1), (27.2)

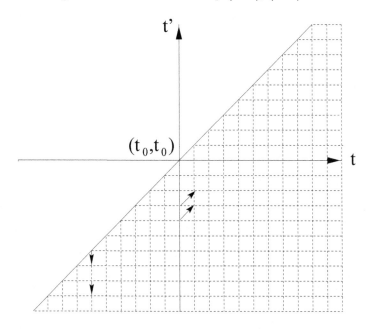

Fig. 27.1   The two-time plane and the numerical strategy.

In the weak coupling case $\alpha = 0.01$ the results do not differ essentially from those obtained within the generalized Kadanoff-Baym approximation. However for an intermediate coupling $\alpha = 0.1$, we get significant deviations. The importance of these deviations is shown also by the rather different shape of the experimentally observed DTS (differential transmission) signals [Betz, Göger, Lauberau, Gartner, Bányai, Haug, Ortner, Becker, Leitenstorfer (2001)] in two different semiconductors like GaAs and CdTe characterized by weak, respectively intermediate couplings to LO phonons as it may be seen in Fig.27.2.

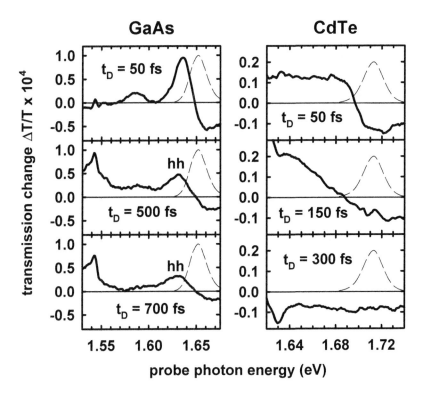

Fig. 27.2   Experimental DTS signal for intermediate and weak coupling, respectively.

The theoretical predictions based on the numerical solution of the quantum kinetic Eqs.(27.1), (27.2) as it can be seen in Fig. 27.3 are indeed in good agreement with differential transmission (DTS) spectra on CdTe [Betz, Göger, Lauberau, Gartner, Bányai, Haug, Ortner, Becker, Leitenstorfer (2001)], while the predictions of a Markovian theory, based on the Boltzmann equation are less satisfactory.

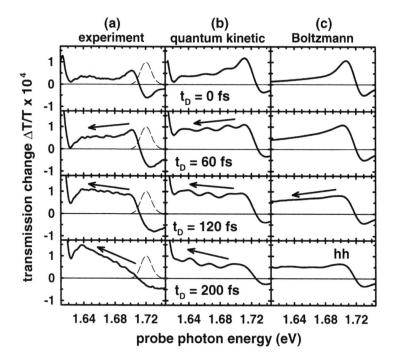

Fig. 27.3 Experimental and theoretical DTS spectra in CdTe taken at different delay times.

Chapter 28

Coulomb quantum kinetics of electrons and holes within the s.c. RPA and the Kadanoff-Baym approximation. Time dependent screening

Now, we want to describe the inclusion of the Coulomb interaction beyond the self-consistent Hartree-Fock scheme, within the frame of the Kadanoff-Baym approximation (like in Chapter 26) in order to have a simple one-time quantum kinetic problem. The inclusion of Coulomb screening effects is important for higher carrier densities created by stronger laser pulses.

To make it simple, we give here the formulae for a system consisting only of electrons, thus omitting band indices, and neglecting the electron-phonon interaction, although in the phenomenological treatments [Hügel, Heinrich, Wegener, Vu, Bányai and Haug (1999)] both were taken into account.

Therefore we consider here only the equation for the electron population

$$
\left.\frac{\partial f_{\vec{k}}(t)}{\partial t}\right|_{coll} = -\int_{-\infty}^{t} dt' \left[\Sigma_{\vec{k}}^{>}(t,t') G_{\vec{k}}^{<}(t',t) - \Sigma_{\vec{k}}^{<}(t,t') G_{\vec{k}}^{>}(t',t) \right.
$$
$$
\left. - G_{\vec{k}}^{>}(t,t') \Sigma_{\vec{k}}^{<}(t',t) + G_{\vec{k}}^{<}(t,t') \Sigma_{\vec{k}}^{>}(t',t) \right] ,
$$

and the simplest self-energy with dressed electron and screened Coulomb lines (see Fig.(28.1)) without vertex correction

$$
\Sigma_{\vec{k}}^{\gtrless}(t,t') = \frac{\imath}{\Omega} \sum_{\vec{q}} G_{\vec{k}-\vec{q}}^{\gtrless}(t,t') W_{\vec{q}}^{\gtrless}(t,t') . \tag{28.1}
$$

Here the the screened Coulomb-potential $W_{\vec{q}}(t,t')$ is defined by the Dyson equations (see Fig.(28.2))

Fig. 28.1   The Fock self-energy with dressed Coulomb line.

$$W_{\vec{q}}^{\gtrless}(t,t') = V_{\vec{q}}\left\{\int_{-\infty}^{t}dt''\,L_{\vec{q}}^{r}(t,t'')W_{\vec{q}}^{\gtrless}(t'',t') + \int_{-\infty}^{t'}dt''\,L_{\vec{q}}^{\gtrless}(t,t'')W_{\vec{q}}^{a}(t'',t')\right\}$$

$$W_{\vec{q}}^{a}(t,t') = V_{\vec{q}}\left\{\delta(t-t'-0) + \int_{t}^{t'}dt''\,L_{\vec{q}}^{a}(t,t'')W_{\vec{q}}^{a}(t'',t')\right\},\qquad(28.2)$$

where

$$W_{\vec{q}}^{a}(t,t') = V_{\vec{q}}\delta(t-t'-0) + \theta(t'-t)[W_{\vec{q}}^{<}(t,t') - W_{\vec{q}}^{>}(t,t')]\,,$$

$$L_{\vec{q}}^{r}(t,t') = \theta(t-t')\left[L_{\vec{q}}^{>}(t,t') - L_{\vec{q}}^{<}(t,t')\right]\,,$$

and V is the bare Coulomb potential

$$V_{\vec{q}} = \frac{4\pi e^2}{\epsilon_0 q^2}\,.$$

Fig. 28.2   The Dyson equation for the dressed Coulomb line $W_{\vec{q}}^{a}$.

On its turn, the polarization bubble $L_{\vec{q}}$ will be taken here without vertex corrections, but with the dressed electron propagators (see Fig.(28.3))

$$L_{\vec{q}}^{\gtrless}(t,t') = -2\frac{\imath}{\Omega\hbar^2}\sum_{\vec{p}}G_{\vec{p}+\vec{q}}^{\gtrless}(t,t')G_{\vec{p}}^{\lessgtr}(t',t).$$

Fig. 28.3   The simplest inter-band polarization bubble with dressed electron lines.

The electron Green functions obey the property

$$G_{\vec{p}}^{\lessgtr}(t,t')^* = -G_{\vec{p}}^{\lessgtr}(t',t),$$

while the polarization bubble satisfies

$$L_{\vec{q}}^{\lessgtr}(t,t')^* = -L_{\vec{q}}^{\lessgtr}(t',t),$$

$$L_{\vec{q}}^{>}(t,t') = L_{-\vec{q}}^{<}(t',t).$$

This last property is more general, and stems from the invariance of the interaction with respect to particle permutation.

We may also show the relation

$$W_{\vec{q}}^{\gtrless}(t,t') = \int_{-\infty}^{t} dt_1 \int_{-\infty}^{t'} dt_2 W_{\vec{q}}^{r}(t,t_1)L_{\vec{q}}^{\gtrless}(t_1,t_2)W_{\vec{q}}^{a}(t_2,t'),$$

and as a consequence analogous properties also for $W_{\vec{q}}$

$$W_{\vec{q}}^{\lessgtr}(t,t')^* = -W_{\vec{q}}^{\lessgtr}(t',t)$$

and

$$W_{\vec{q}}^{>}(t,t') = W_{-\vec{q}}^{<}(t',t). \tag{28.3}$$

These both properties ensure the reality of the collision term and the conservation of the particle number in the collision term.

Now we want to implement the generalized Kadanoff-Baym relations

$$G_{\vec{p}}^{<}(t,t') \doteq -G_{0\vec{p}}^{r}(t,t')f_{\vec{p}}(t') + G_{0\vec{p}}^{a}(t,t')f_{\vec{p}}(t)\,,$$

$$G_{\vec{p}}^{>}(t,t') \doteq G_{0\vec{p}}^{r}(t,t')(1 - f_{\vec{p}}(t')) - G_{0\vec{p}}^{a}(t,t')(1 - f_{\vec{p}}(t))\,,$$

in order to close the equation for the population. Here we take the free retarded and advanced Green functions

$$G_{0\vec{p}}^{r}(t,t') = G_{0\vec{p}}^{a}(t',t)^{*} = -\imath\theta(t - t')e^{-\frac{\imath}{\hbar}\epsilon_{\vec{p}}^{0}(t-t')}\,.$$

Since the free-particle energies $\epsilon_{\vec{p}}^{0}$ depend only on $|\vec{p}|$ it follows that all the entities $G_{\vec{p}}, f_{\vec{p}}, L_{\vec{p}}, W_{\vec{p}}$ are functions of only the absolute value $|\vec{p}|$. Therefore, the collision term may be written as:

$$\frac{\partial f_k(t)}{\partial t}\bigg|_{coll} = \frac{2}{(4\pi)^2 k} \int_0^\infty dk'^2 \int_0^\infty dq^2 \int_{-\infty}^t dt'\theta(1 - |\frac{k^2 + k'^2 - q^2}{2kk'}|)$$

$$\times \Im\left[f_k(t')(1 - f_{k'}(t'))e^{\frac{\imath}{\hbar}(\epsilon_k^0 - \epsilon_{k'}^0)(t-t')}W_q^{>}(t,t') - (k \rightleftharpoons k')\right]\,.$$

Using the symmetry relations, the closed equation for the screened Coulomb potential is

$$W_q^{>}(t,t') = V_q\left\{L_q^{>}(t,t')V_q + \int_{-\infty}^t dt'' L_q^{r}(t,t'')W_q^{>}(t'',t')\right.$$

$$\left. + \int_{-\infty}^{t'} dt'' L_q^{>}(t,t'') \left[W_q^{>}(t',t'') - W_q^{>}(t'',t')\right]\right\}\,.$$

Using also the explicit expression for the inter-band polarization bubble, we get

$$L_q^{r}(t,t'') = L_q^{>}(t,t'') - L_q^{>}(t'',t) = 2\Re L_q^{>}(t,t'') \qquad \text{for} \qquad t > t''\,,$$

with

$$L_q^{>}(t,t'') = -2\imath\frac{1}{\hbar^2(4\pi)^2 q} \int_0^\infty dp'^2 \int_0^\infty dp^2\theta(1 - |\frac{q^2 + p'^2 - p^2}{2qp'}|)$$

$$\times e^{\frac{\imath}{\hbar}(\epsilon_p^0 - \epsilon_{p'}^0)(t-t'')} [f_p(t'')(1 - f_{p'}(t''))\theta(t - t'') + f_p(t)(1 - f_{p'}(t))\theta(t'' - t)]\,.$$

The equation for $W_{\vec{q}}^{>}(t,t')$ may also be rewritten in a closed form involving only the screened potential with the second time less than the first:

$$W_q^{>}(t,t') = V_q \{ L_q^{>}(t,t') V_q + \int_{-\infty}^{t} dt'' L_q^r(t,t'') \left[W_q^{>}(t'',t')\theta(t''-t') \right]$$

$$- W_q^{>}(t',t'')^{*}\theta(t'-t'') \right] + 2\int_{-\infty}^{t'} dt'' L_q^{>}(t,t'')\Re W_q^{>}(t',t'') \}.$$

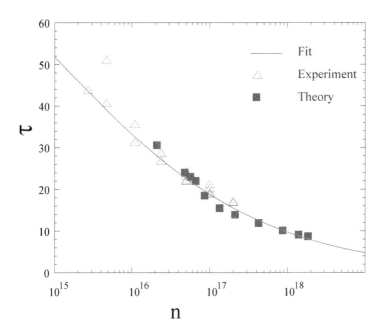

Fig. 28.4   Decay time τ of the time-integrated FWM signal versus carrier density n.

In order to make phenomenological predictions, one has to consider both electrons and holes, and one has to calculate not only the populations, but also the inter-band polarization. Even more, the interaction with the LO-phonons has to be taken also into account (at least on the same level as in Chapter 26). The presence of both Coulomb and phonon lines in the theory manifests itself first in the existence of two different self-energy diagrams and in a modification of the Dyson equation for the screened Coulomb potential, since the inter-band polarization bubble may mix the Coulomb line with the LO-phonon line. The theoretical predictions for FWM-experiments at higher carrier densities based on the above described scheme [Hügel, Heinrich, Wegener, Vu, Bányai and Haug (1999)] are very encouraging. In Fig.28.4 one may see the measured and the predicted decay time of the integrated FWM signal versus carrier density n_{eh}.

Chapter 29

Full two-time quantum kinetics of an electron-hole Coulomb system. Ward identities

In this chapter we analyze the problems encountered in the formulation of a full two-times quantum kinetics, in the case of the (screened) Coulomb interaction between the particles. Unfortunately, such an extension - beyond the Kadanoff-Baym Ansatz - is not at all trivial, due to the presence of infrared singularities if the Ward identity is violated. These are not the intrinsic infrared divergences of the Coulomb theory, but spurious ones, introduced by the approximations.

We consider here the conduction and valence band problem and a Coulomb self-energy without vertex corrections

$$\Sigma_{\vec{k},\mu\nu}(t,t') = \frac{\imath}{\Omega} \sum_{\vec{q}} W_{\vec{q}}(t,t') G_{\vec{k}-\vec{q},\mu\nu}(t,t') \,. \tag{29.1}$$

The screened Coulomb potential is given as before by

$$W_{\vec{q}}^{r}(t,t') = V_{\vec{q}}\delta(t,t') + V_{\vec{q}}L_{\vec{q}}^{r}(t,t_1)W_{\vec{q}}^{r}(t_1,t') \,, \tag{29.2}$$

$$W_{\vec{q}}^{a}(t,t') = W_{\vec{q}}^{r}(t',t)^{*} \,, \tag{29.3}$$

$$W_{\vec{q}}^{\gtrless}(t,t') = W_{\vec{q}}^{r}(t,t_1)L_{\vec{q}}^{\gtrless}(t_1,t_2)W_{\vec{q}}^{a}(t_2,t') \,. \tag{29.4}$$

with the property

$$W_{\vec{q}}^{>}(t,t') = W_{-\vec{q}}^{<}(t',t) \,.$$

The choice of the inter-band polarization bubble we do not fix yet.

It may be seen [Gartner, Banyai, Haug (2000)] that the theory is free of divergences (at most with integrable singularities at $q \to 0$), if the singularity in the dressed (screened) Coulomb potential is not worse than $1/q^2$, because in this case the 3D integral over $\vec{q}$ defining the self-energy

Eq.(29.1) exists (just like in the case of the bare Coulomb interaction). From the equation of the advanced (or retarded) dressed Coulomb line ("screened Coulomb potential") Eq.(29.4) it is clear that - unless the retarded loop vanishes as q^2 for $q \to 0$ - the singularity induced by the bare potential becomes higher and higher with each iteration. On the other hand, if the retarded inter-band polarization loop does vanish as q^2, then $W_{\vec{q}}^r$ (and $W_{\vec{q}}^a$ too) has only the $1/q^2$ singularity inherited from the bare interaction.

It is obvious from this argument that one must have

$$L_{\vec{q}}^{r,a}(t,t') \sim q^2 \; for \; q \to 0 \,.$$

Even if this is the case, the singularities of $W_{\vec{q}}^r$ and $W_{\vec{q}}^a$ in the expression of the greater or lesser component $W_{\vec{q}}^>$ and $W_{\vec{q}}^<$ respectively add up to $1/q^4$. As stated above, this is too high for the theory to be well-defined, and a compensating behavior is needed from the corresponding bubble component

$$L_{\vec{q}}^{\gtrless}(t,t') \sim q^2 \; for \; q \to 0 \,.$$

Therefore, we can see that the quadratic vanishing is required for all the components of the inter-band polarization loop.

In the exact theory this condition is automatically ensured by the Ward identity, expressing the particle number conservation:

$$\frac{\partial}{\partial t'} \sum_{\vec{p}\nu} a_{\vec{p}\nu}^+(t') a_{\vec{p}\nu}(t') = 0 \,.$$

(The index ν here runs over the conduction and valence bands.) One readily proves, like in the quantum field theory [Weinberg (1996)]

$$\frac{\partial}{\partial t'} \sum_{\vec{p}\nu} \frac{1}{i^2} \left\langle T_c \left[a_{\vec{k}\mu_1}(t_1) a_{\vec{k}\mu_2}^+(t_2) a_{\vec{p}\nu}^+(t') a_{\vec{p}\nu}(t') \right] \right\rangle$$

$$= \frac{1}{i} \left[\delta_c(t'-t_2) - \delta_c(t'-t_1) \right] G_{\vec{k},\mu_1\mu_2}(t_1,t_2) \,. \qquad (29.5)$$

On the LHS of Eq. (29.5) one has a two-particle Green function. The δ-terms appearing on the RHS stem from the discontinuities implied by the contour-ordering operator. In the diagrammatic expansion the two-particle Green function has in general four external points. Here two of them are joined into one, with coordinates $(\vec{p}, t')$. The expansion contains a disconnected part corresponding to the contraction

$$\left\langle T_c \left[a_{\vec{k}\mu}(t_1) a^+_{\vec{k}\mu_2}(t_2) \right] \right\rangle \sum_{\vec{p}\nu} \left\langle a^+_{\vec{p}\nu}(t') a_{\vec{p}\nu}(t') \right\rangle , \tag{29.6}$$

all the other diagrams forming the connected part.

Two things are worth mentioning here: In the first place, the disconnected part has a vanishing t'-derivative, as seen immediately from the particle number conservation Eq.(23.2). Therefore Eq. (29.5) is also obeyed, if on its LHS we take only the connected part of the two-particle Green function. In the second place, if one joins in the connected part the other two external points $(t_1 = t_2 = t, \quad \mu_1 = \mu_2 = \mu)$ and sums over $\vec{k}$ and μ, one obtains the interband polarization loop at $\vec{q} = 0$. Therefore one has

$$\frac{\partial}{\partial t'} L_0(t, t') = 0 . \tag{29.7}$$

By a similar argument, the t-derivative is shown to vanish as well. On the other hand, before the arrival of the optical pulse the inter-band polarization loop is identically zero (there is no screening in the absence of carriers); therefore, it is clear that the $\vec{q} = 0$ component of the loop is always zero. Due to the spherical symmetry of the problem, the vanishing is independent of the direction of $\vec{q}$. On the other hand, by arguments of analyticity in $\vec{q}$ (implicit in the perturbation theory), one gets a quadratic behavior at small momenta, and this is the desired result.

Eq.(29.5) can be cast in a form which is closer to that used in the equilibrium many-body theory. The connected part of the two-particle Green function of Eq.(29.5) can be expressed as a product of two one-particle Green functions and the vertex part. Moving the Green functions to the RHS, one is left with

$$\frac{\partial}{\partial t'} \Gamma_{\vec{p},\mu_1;0;\vec{p},\mu_2}(t_1, t', t_2) = \imath \left(\delta_c(t' - t_1) - \delta_c(t' - t_2) \right) G^{-1}_{\vec{p},\mu_1\mu_2}(t_1, t_2) . \tag{29.8}$$

This is a relation between the vertex part and the self-energy which - in the particular case of equilibrium and in the frequency domain - reduces to the well-known connection between the vertex part and the frequency derivative of the self-energy. Remember that the Ward identity in the T=0 theory looked (after Fourier transform in the time variables) as

$$\Gamma^{\alpha\alpha'}_{p0p} = -\imath e \frac{\partial}{\partial p_0} G^{-1}(p)^{\alpha\alpha'} .$$

The conditions of a divergence-free theory are thus ensured by the Ward identity in the exact theory. In approximate schemes they cannot be taken for granted, and an analysis of the specific situation is needed.

Let us consider first the problem with a bubble diagram without vertex part

$$L_0^r(t,t') = \theta(t-t')\frac{-2\imath}{\Omega}\sum_{\vec{p}}\sum_{\mu\nu}\Big[G^>_{\vec{p},\mu\nu}(t,t')G^<_{\vec{p},\nu\mu}(t',t) \qquad (29.9)$$

$$- \, G^<_{\vec{p},\mu\nu}(t,t')G^>_{\vec{p},\nu\mu}(t',t)\Big] \, ,$$

within the Generalized Kadanoff-Baym Ansatz

$$G^{\gtrless}_{\vec{p},\mu\nu}(t,t') \approx \sum_{\lambda}\Big[\imath G^r_{\vec{p},\mu\lambda}(t,t')G^{\gtrless}_{\vec{p},\lambda\nu}(t',t') - \imath G^{\gtrless}_{\vec{p},\mu\lambda}(t,t)G^a_{\vec{p},\lambda\nu}(t,t')\Big] \, ,$$

$$(29.10)$$

as we considered it in the previous Chapter.

Due to this approximation, the divergences of the retarded and advanced loops are easily shown to vanish for $\vec{q}=0$. Indeed, inserting the Kadanoff-Baym approximation one gets

$$L_0^r(t,t') \approx \theta(t-t')\frac{-2\imath}{\Omega}\sum_{\vec{p}}\sum_{\mu\nu\lambda\lambda'} G^r_{\vec{p},\mu\lambda}(t,t')\Big\{G^>_{\vec{p},\lambda\nu}(t',t')G^<_{\vec{p},\nu\lambda'}(t',t')$$

$$- \, G^<_{\vec{p},\lambda\nu}(t',t')G^>_{\vec{p},\nu\lambda'}(t',t')\Big\} G^a_{\vec{p},\lambda'\mu}(t',t) \, . \qquad (29.11)$$

But at equal times $G^>_{\vec{p}}$ and $G^<_{\vec{p}}$ commute, so that the terms in the curly brackets cancel out. A similar result holds for L^a, and this implies the desired $1/q^2$ behavior for $W^{r,a}_{\vec{q}}$.

The argument shows that differences between $L^>$ and $L^<$ are well-behaved around $\vec{q}=0$, but not $L^>$ and $L^<$ themselves. This still leaves a strong $1/q^4$ singularity for $W^{\gtrless}_{\vec{q}}$, which renders the self-energy ill-defined.

Fortunately, the kinetic equation for equal-time Green functions contains the self-energy Σ in combinations in which the leading singularity is canceled out. We illustrate this cancellation effect in the case of the collision term appearing in the kinetic equation for $G^<(t,t)$. It reads

$$
\frac{\partial}{\partial t} G^<_{\vec{k},\mu\nu}(t,t)\bigg|_{coll} = \frac{1}{\Omega} \sum_{\vec{q}} \sum_{\lambda} \int_{-\infty}^{t} dt' \left\{ G^>_{\vec{k}-\vec{q},\mu\lambda}(t,t') W^>_{\vec{q}}(t,t') G^<_{\vec{k},\lambda\nu}(t',t) \right.
$$

$$
- G^<_{\vec{k}-\vec{q},\mu\lambda}(t,t') W^<_{\vec{q}}(t,t') G^>_{\vec{k},\lambda\nu}(t',t)
$$

$$
- G^>_{\vec{k},\mu\lambda}(t,t') W^<_{\vec{q}}(t',t) G^<_{\vec{k}-\vec{q},\lambda\nu}(t',t)
$$

$$
\left. + G^<_{\vec{k},\mu\lambda}(t,t') W^>_{\vec{q}}(t,t') G^>_{\vec{k}-\vec{q},\lambda\nu}(t',t) \right\} .
$$

Using the symmetry property Eq.(28.3), it is clear that for $\vec{q} \to 0$ the first and the second term are compensated by the third and, respectively, the fourth one. Therefore, the leading singularity cancels out, and only integrable singularities remain.

In a true two-time version of the theory, the above described simplifications do not occur, and one has to treat again the problem of $q \to 0$ divergences within an approximate theory, starting from the simple self-energy without vertex correction Eq. (29.1)

Baym and Kadanoff [Baym, Kadanoff (1961); Baym (1962)] gave a recipe to get a charge conserving s.c. approximation. It implies the choice of a self-energy as a functional of the propagator $\Sigma(G)$, and respectively an inter-band polarization bubble determined by the Bethe-Salpeter equation in which the effective interaction (vertex) is given by $\partial\Sigma/\partial G$. The situation is recursive: the self-energy itself is determined by the vertex part, yet no convergence seems possible. Simple $\Sigma(G)$ functionals give rise to complicated vertex parts, in disagreement with the vertex originally taken in Σ.

A way out of this difficulty is obtained by noting that the proof of the Baym-Kadanoff result still holds under simplified circumstances. Instead of taking the full functional derivative $\partial\Sigma/\partial G$, one takes the derivative only with respect to the Green functions which form the path connecting the external points of the self-energy.

In this case the modified procedure is a drastic simplification, because one has only one Green function connecting the two endpoints of Σ. Besides, the vertex part - not being involved in the functional derivative any more - the recursive chain is broken.

It can be shown that this recipe leads to a non-trivial vertex part in the Coulomb bubble, namely, the one given by the solution of the ladder equation Fig.(23.2)

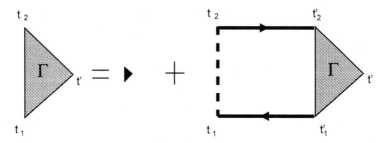

Fig. 29.1   The summation of the ladder diagrams for the vertex part.

To summarize, we start now again from the Dyson equations for the particle Green functions and for the dressed Coulomb line with the s.c. RPA particle self-energy without vertex correction Eq.(28.1). However, we do not use the generalized Kadanoff-Baym Ansatz, and try to construct a true two-time Coulomb quantum kinetics.

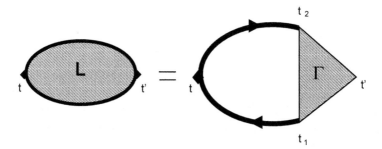

Fig. 29.2   Insertion of the vertex part into the inter-band polarization bubble.

Keeping the vertex part Γ in the inter-band polarization bubble (see Fig.(23.2)) one has

$$L_{\vec{q}}(t,t') = -\frac{2i}{\Omega}\sum_{\vec{p}}\int_c dt_1 \int_c dt_2\, G_{\vec{p}+\vec{q},\mu\lambda_1}(t,t_1)\Gamma_{\vec{p}+\vec{q},\lambda_1;\vec{q};\vec{p},\lambda_2}(t_1,t',t_2)G_{\vec{p},\lambda_1\mu}(t_2,t),$$

where the integrations are performed on the Keldysh contour. The vertex part on its turn is then determined by the ladder equation

$$\Gamma_{\vec{p}+\vec{q},\lambda_1;\vec{q},\vec{p},\lambda_2}(t_1,t',t_2) = \delta(t_1-t')\delta(t'-t_2)\delta_{\lambda_1,\lambda_2}$$

$$+ \frac{\imath}{\Omega}\sum_{\vec{l}} W_{\vec{l}}(t_2,t_1)\int_c dt'_1\int_c dt'_2 G_{\vec{p}+\vec{q}+\vec{l},\lambda_1,\lambda'_1}(t_1,t'_1)\Gamma_{\vec{p}+\vec{q}+\vec{l},\lambda'_1;\vec{q},\vec{p}+\vec{l},\lambda'_2}(t'_1,t',t'_2)$$

$$\times\, G_{\vec{p}+\vec{l},\lambda'_2,\lambda_2}(t'_2,t_2)\,. \tag{29.12}$$

This construction ensures that

$$\frac{\partial}{\partial t}\lim_{q\to 0} L_{\vec{q}}(t,t') = \frac{\partial}{\partial t'}\lim_{q\to 0} L_{\vec{q}}(t,t') = 0\,,$$

which, as above described, implies also $\lim_{q\to 0} L_{\vec{q}}(t,t') = 0$.

A major problem for a numerical simulation is, however, the number of time variables - which the computers at our disposal cannot yet handle.

In these last few Chapters we gave a presentation of quantum kinetics within the diagrammatic Keldysh-Green-function formalism based on the summation of certain diagrams for the Green functions. This leads to non-local, non-linear integral equations. I'd like to mention that this is by no means the only approach to quantum kinetics. As we have seen on some examples along these lectures, an alternative is to consider the equations of motion for the truncated hierarchy of averages at equal times. This last version is mostly suitable for coherent problems, where the peculiarities of the dissipation are less important. Generally speaking, the two ways are not equivalent, especially with regard to the initial conditions one has to give. Both methods give rise to complicated equations whose mathematical properties are not well understood, and for which only numerical solutions are available. Nevertheless, along both ways remarkable successes were achieved in the interpretation of experimental data on ultra-short-fast spectroscopy. That is the reason for the inclusion of experimental references only in this part of the lectures devoted to the most recent developments.

Appendix A

H-theorem for bosons

For the following discussion it is useful to define the symmetrized transition rates

$$W_{\vec{k}\vec{k}'}e^{\beta e_{\vec{k}'}} = W_{\vec{k}'\vec{k}}e^{\beta e_{\vec{k}}} = \mathcal{W}_{\vec{k}\vec{k}'} = \mathcal{W}_{\vec{k}'\vec{k}} \geq 0. \tag{A.1}$$

Properties of the rate equation in a finite volume:

The properties of the rate equation stem from the special 'gain-loss' structure of the equation, from the positivity of the transition rates and from another property of the rates known as connectivity. In order to explain the latter, we can regard the set of the states $\{\vec{k}\}$ as split into connectivity classes defined as follows: $\vec{k}$ and $\vec{k}'$ belong to the same class if one can reach the state $\vec{k}'$ from the state $\vec{k}$ by a finite number of allowed transitions. This means there is a sequence $\vec{k}_1, \vec{k}_2, ...\vec{k}_n$, so that $\mathcal{W}_{\vec{k}\vec{k}_1}, \mathcal{W}_{\vec{k}_1\vec{k}_2}...\mathcal{W}_{\vec{k}_n\vec{k}'}$ are all non-vanishing. It is clear that states belonging to different classes evolve separately, without influencing each other (separate fluids). We will assume the following connectivity property of the transition rates: *the states $\{\vec{k}\}$ form one connectivity class.* In other words, any state can be reached starting from any state.

We are now in a position to formulate the above-mentioned properties of the solution $f_{\vec{k}} \equiv \langle a_{\vec{k}}^{+}a_{\vec{k}}\rangle$ of Eq. (21.7) :

a) *Positivity*: If the initial condition is positive $f_{\vec{k}}(0) \geq 0$, then $f_{\vec{k}}$ stays positive for any later time. For the proof consider the situation in which some $f_{\vec{k}}$ components are vanishing, while the others are strictly positive. For the derivative of the vanishing components the rate equation gives

$$\frac{\partial}{\partial t}f_{\vec{k}} = \frac{1}{V}\sum_{\vec{k}'}\mathcal{W}_{\vec{k}\vec{k}'}f_{\vec{k}'}e^{-\beta e_{\vec{k}}} \geq 0. \tag{A.2}$$

If the derivative is strictly positive, the corresponding component is strictly increasing, and becomes strictly positive. On the one hand we are left with discussion of the case where the derivative is zero too, and the result depends on the sign of higher derivatives. On the other hand, the above equation shows that $\frac{\partial}{\partial t} f_{\vec{k}} = 0$, if and only if $f_{\vec{k}'} = 0$ for any $\vec{k}'$ for which $\mathcal{W}_{\vec{k}\vec{k}'} \neq 0$. In other words, for a vanishing $f_{\vec{k}}$ component, the vanishing of its first derivative is equivalent to the vanishing of the components which are one jump apart. In this situation the second derivative is:

$$\frac{\partial^2}{\partial t^2} f_{\vec{k}} = \frac{1}{V^2} \sum_{\vec{k}'\vec{k}''} \mathcal{W}_{\vec{k}\vec{k}'} \mathcal{W}_{\vec{k}'\vec{k}''} f_{\vec{k}''} e^{-\beta(e_{\vec{k}}+e_{\vec{k}'})} \geq 0. \tag{A.3}$$

If the second derivative is vanishing, the components two jumps away are also vanishing. The argument goes on recursively, until a strictly positive component is reached. This is bound to happen due to the connectivity property. If this occurs after n steps, the first $n-1$ derivatives are zero and the n-th is strictly positive. This concludes the proof of our statement.

Note that we have shown in fact a stronger result, namely that for $t > 0$, one has $f_{\vec{k}}(t) > 0$ for any $\vec{k}$.

b) Convergence to equilibrium: In order to show that the rate equation Eq.(21.7) describes the irreversible relaxation to the equilibrium Bose Einstein distribution, one makes use of the famous H-theorem.

The H-function is defined as

$$H = \frac{1}{V} \sum_{\vec{k}} \left[(1 + f_{\vec{k}}) ln(1 + f_{\vec{k}}) - f_{\vec{k}} ln f_{\vec{k}} - \beta e_{\vec{k}} f_{\vec{k}} \right], \tag{A.4}$$

and makes sense even if some $f_{\vec{k}} = 0$, because one may define by continuity $x \ln x = 0$ for $x = 0$.

The time derivative of the H-function is given by

$$\frac{\partial}{\partial t} H = \frac{1}{V} \sum_{\vec{k}} \left[\ln(1 + f_{\vec{k}}) - \ln f_{\vec{k}} - \beta e_{\vec{k}} \right] \frac{\partial}{\partial t} f_{\vec{k}}$$

$$= \frac{1}{2V^2} \sum_{\vec{k}\vec{k}'} \mathcal{W}_{\vec{k}\vec{k}'} \left\{ \ln \left[f_{\vec{k}}(1 + f_{\vec{k}'}) e^{-\beta e_{\vec{k}'}} \right] - \ln \left[f_{\vec{k}'}(1 + f_{\vec{k}}) e^{-\beta e_{\vec{k}} n} \right] \right\}$$

$$\times \left[f_{\vec{k}}(1 + f_{\vec{k}'}) e^{-\beta e_{\vec{k}'}} - f_{\vec{k}'}(1 + f_{\vec{k}'}) e^{-\beta e_{\vec{k}}} \right]. \tag{A.5}$$

In this derivation the symmetry of $\mathcal{W}_{\vec{k}\vec{k}'}$ was used. Since the logarithmic function is increasing, the difference between the two logarithms has the same sign as the difference between their arguments. Therefore one has the following statement, known as the H-theorem:

$$\frac{\partial}{\partial t}H \geq 0, \tag{A.6}$$

which says that H is an increasing function of time. Next, we show that H is bounded from above, and therefore it has a finite limit at $t \to \infty$. First, note that there exist $a, b > 0$, so that

$$h(x) = (x+1)ln(x+1) - xlnx < ax + b \qquad \text{for} \quad \text{any} \quad x \geq 0. \tag{A.7}$$

Indeed, for large x one has the result

$$\lim_{x \to \infty} \frac{h(x)}{x} = 0, \tag{A.8}$$

implying that $h(x)$ grows slower than any linear function, so we can choose any $a > 0$ so that $h(x) - ax$ is bounded from above by some positive constant b.

Using this inequality one may write

$$H \leq \frac{1}{V} \sum_{\vec{k}} [af_{\vec{k}} + b - \beta e_{\vec{k}} f_{\vec{k}}] \leq an_{tot} + \frac{b}{V}N, \tag{A.9}$$

where N is the total number of the states $\{\vec{k}\}$. In order to simplify the proof, we assume this number to be finite, i.e. we impose an upper cut-off in energy. We can see now that the H-function is smaller than a time-independent quantity.

The conclusion is that in the limit $t \to \infty$ a stationary state is reached, for which the inequality in the H-theorem becomes a strict equality. This entails the vanishing of all the terms in the sum giving $\partial/\partial t H$ and thus

$$f_{\vec{k}}(1 + f_{\vec{k}'})e^{-\beta e_{\vec{k}'}} = f_{\vec{k}'}(1 + f_{\vec{k}})e^{-\beta e_{\vec{k}}}, \qquad \text{if} \qquad W_{\vec{k}\vec{k}'} \neq 0. \tag{A.10}$$

In other words the quantity

$$\frac{(1 + f_{\vec{k}})e^{-\beta e_{\vec{k}}}}{f_{\vec{k}}} = \alpha \tag{A.11}$$

is constant, by the connectivity property for all the states of the system. This is equivalent to

$$f_{\vec{k}} = \frac{1}{e^{\beta(e_{\vec{k}} - \mu)} - 1} = f^0(e_{\vec{k}}, \mu), \tag{A.12}$$

where f^0 is the Bose function and μ is defined by $\alpha = e^{-\beta\mu}$.

Properties of the rate equation Eq. (21.11) in an infinite volume: The same properties hold in this case too:

a) *Positivity:* is shown by a similar argument. A new feature is the fact - obvious by the examination of the rate equation Eq.(21.11) - that if $n_0(0) = 0$ then $n_0(t) = 0$ for any $t > 0$.

b) *Convergence to equilibrium:* This is proven along the same lines. The infinite volume version of the H-function is

$$H = \int \frac{d^3k}{(2\pi)^3} \left[(1 + f_{\vec{k}}) ln(1 + f_{\vec{k}}) - f_{\vec{k}} ln f_{\vec{k}} - \beta e_{\vec{k}} f_{\vec{k}} \right], \tag{A.13}$$

with no contribution from the condensate. This stems from the fact that

$$\frac{1}{V} h(f_0) = \frac{f_0}{V} \cdot \frac{h(f_0)}{f_0} \tag{A.14}$$

goes to zero in the limit $f_0 \to \infty$, $V \to \infty$, $f_0/V = n_0$.

In order to establish the H-theorem, we use the rate equation to calculate $\frac{\partial}{\partial t} H$ and obtain

$$\frac{\partial}{\partial t} H = \frac{1}{2} \int \frac{d^3k}{(2\pi)^3} \frac{d^3k'}{(2\pi)^3} \mathcal{W}_{\vec{k}\vec{k}'} \left\{ ln \left[f_{\vec{k}} (1 + f(\vec{k}')) e^{-\beta e_{\vec{k}'}} \right] \right. \tag{A.15}$$

$$\left. - ln \left[f(\vec{k}')(1 + f_{\vec{k}}) e^{-\beta e_{\vec{k}}} \right] \right\} \left[f_{\vec{k}} (1 + f(\vec{k}')) e^{-\beta e_{\vec{k}'}} - f(\vec{k}')(1 + f_{\vec{k}}) e^{-\beta e_{\vec{k}}} \right]$$

$$+ n_0 \int \frac{d^3k}{(2\pi)^3} \mathcal{W}_{\vec{k}0} \left\{ ln f_{\vec{k}} - ln \left[(1 + f_{\vec{k}}) e^{-\beta e_{\vec{k}}} \right] \right\} \left[f_{\vec{k}} - (1 + f_{\vec{k}}) e^{-\beta e_{\vec{k}}} \right], \tag{A.16}$$

which is positive, by the same argument. The upper bound for H is established in the same way, under the assumption of a cut-off spectrum. At $t \to \infty$ one reaches a stationary state, for which $\partial/\partial t H = 0$. It is known that positive continuous functions can have vanishing integrals only if they are identically zero. This leads, as in the discrete case, to

$$\frac{(1 + f_{\vec{k}})e^{-\beta e_{\vec{k}}}}{f_{\vec{k}}} = \alpha \,, \tag{A.17}$$

with the additional feature that if in the final state $n_0 \neq 0$ then $\alpha = 1$ ($\mu = 0$).

In order to classify the possible scenarios, we can remark that for any $\mu \leq 0$ one has the inequality

$$\int \frac{d^3 k}{(2\pi)^3} f^0(e_{\vec{k}}, \mu) \leq \int \frac{d^3 k}{(2\pi)^3} f^0(e_{\vec{k}}, 0) = n_{cr} \,. \tag{A.18}$$

Therefore, initial conditions with a density n_{tot} smaller than the critical density n_{cr} can be accommodated only if the final condensate is zero. The corresponding equilibrium situation is

$$n_0(\infty) = 0 \,, \tag{A.19}$$

$$f_{\vec{k}}(\infty) = f^0(e_{\vec{k}}, \mu) \,, \tag{A.20}$$

with μ given by the average particle density

$$n_{tot} = \int \frac{d^3 k}{(2\pi)^3} f^0(e_{\vec{k}}, \mu) \,. \tag{A.21}$$

If $n_{tot} > n_{cr}$ the final state should be

$$f_{\vec{k}}(\infty) = f^0(e_{\vec{k}}, 0) \,, \tag{A.22}$$

$$n_0(\infty) = n_{tot} - \int \frac{d^3 k}{(2\pi)^3} f^0(e_{\vec{k}}, 0) \,. \tag{A.23}$$

Bibliography

Abrikosov, A. A., Gorkov, L. P. and Dzialoshinski, J. P. (1965). Quantum field theoretical methods in statistical physics, *Pergamon Press*, pp. 1–365

Adler, S. (1962). Quantum theory of the dielectric Constant in Real Solids. *Phys. Rev.* **126**, 2, pp 413–420.

Anderson, M. H., Ensher, J. R., Matthews, M.R., Wiemann, C. E. and Cornell, E. A. (1995). Observation of Bose-Einstein condensation in a dilute atomic vapor. *Science* **269**, 5221 , pp. 198–201.

Argyres, P. N. (1960).. Quantum Transport in a Magnetic Field, *Phys. Rev.* **117**, 2, pp 315–328.

Bányai, L. (1969). Teoria cuantica a coeficientilor de transport, *Seminar de Fizica Teoretica, Ed. V. Novacu, Bucuresti* **1**, pp. 13–35.

Bányai, L. and Aldea, A. (1979). Master-Equation Approach to the Hopping Transport Theory, *Fortschritte der Physik* **27**, 9, pp. 435–462.

Bányai, L. and Gartner, P. (1984). Clausius-Mosotti limit of the quantum theory of the electronic dielectric constant, *Phys. Rev.* **B29**, 2, pp. 728–734.

Bányai, L. (1984). The scaling relation between the microscopic (kinetic) and macroscopic (hydrodynamic) descriptions, *Frontiers of Nonequilibrium Statistical Physics. Eds. G.T. Moore and M. O. Scully NATO ASI Series , Santa Fe. Plenum Press* **B135**, pp. 241–257.

Bányai, L., Aldea, A. and Gartner, P. (1984). On the Nyquist noise, *Z. Phys. B - Condensed Matter* **58**, pp. 161–164

Bányai, L. (1993). Motion of a Classical Polaron in a dc Electric Field, *Phys. Rev. Lett.* **70**, 11, pp. 1674–1677.

Bányai, L. , Tran Thoai, D.B., Reitsamer, E., Haug, H., Steinbach, D., Wehner, M.U., Wegener, M. , Marschner, T. and Stolz, W. (1995). Exciton-LO-phonon quantum kinetics: evidence of memory effects in bulk GaAs. *Phys. Rev. Lett.* **75**, 11, pp. 2188–2191.

Bányai, L., Gartner, P., Schmitt, O.M., and Haug,.H. (2000). Condensation kinetics for bosonic excitons interacting with a thermal phonon bath, *Phys. Rev.* **B61**, 13, pp. 8823–8834

Bányai, L. and Gartner, P. (2002). Real time Bose-Einstein condensation in a finite volume with a discrete spectrum, *Phys. Rev. Letters* **88**, 21, pp. 210404-

1–210404-4.

Baym, G., Kadanoff, L.P. (1961). Conservation laws and correlation functions. *Phys. Rev.* **124**, 2, pp. 287–299.

Baym, G. (1962). Self-consistent approximationsin many-body systems. *Phys. Rev.* **127**, 4, pp. 1391–1401 .

Betz, M., Göger, G., Lauberau, A., Gartner, P., Bányai, L., Haug, H., Ortner, K., Becker, C.R., Leitenstorfer, A. (2001). Subthreshold carrier-LO phonon dynamics in semiconductors with intermediate polaron coupling:A purely quantum kinetic relaxation channel. *Phys. Rev. Lett.* **86**, 20, pp. 4684–4687

Bonitz, M. (1998) Quantum Kinetic Theory, *B. G. Teubner Stuttgart, Leipzig* pp. 1–388.

Brenig, W. (1989) Statistical Theory of Heat-Nonequilibrium Phenomena, *Springer, Berlin* pp. 1– .

Caldeira, A. O. and Leggett, A. J. (1983). Quantum tunnelling in a dissipative system. *Ann. Phys.* **149**, 2, pp. 374–456.

Cohen-Tanoudji, C. (1992). Atom-photon Interactions, *Wiley*, pp. 1–656

Davies, E.B. (1967). Quantum theory of open systems, *Academic Press, London* pp. 1–171.

Davydov, I. and Pomeranchuk, I. (1940) *J. Phys. USSR* **2**, 2 , pp. 147–160.

Davis, K.B., Mewes, M.O., Andrews, M.R., van Druten, N.J., Durfee, S., Kurn, D.M., and Ketterle, W. (1995) Bose-Einstein condensation in a gas of sodium atoms. *Phys. Rev. Lett.* **75**, 22, pp 3969–3973.

Gartner, P., Bányai, L. and Haug, H. (1999). Two-time electron-LO-phonon quantum kinetics and the generalized Kadanoff-Baym approximation. *Phys. Rev.* **B60**, 20, pp. 14234–14241.

Gartner, P., Bányai, L. and Haug, H. (2000). Coulomb screening in the two-time Keldysh-Green-function formalism. *Phys. Rev.* **B62**, 11, pp. 7116–7120.

Gorini, V., Frigerio, A., Verry, M., Kossakowski, A. and Sudarshan, E. C. G. (1978). Properties of quantum markovian master equations, *Rep. Math. Phys. (GB)* **13**, 19, pp. 149–173

Götze, W. and Hajdu, J. (1978). On the theory of the transverse dynamic magneto-conductivity, *J. Phys. C* **11**, 19, pp. 3993–4008

Götze, W. and Hajdu, J. (1979). On the transverse dynamic magneto-conductivity in two-dimensional electron-impurity systems, *Solid State Commmun.* **29**, 2, pp. 89–92.

Greenwood, D. A. (1958). The Boltzmann equation in the theory of electric al conduction in metalls, *Proc. Phys. Soc.* **71**, 4, pp. 585–596.

Haug, H. and Jauho, A. P. (1995). Quantum kinetics for transport and optics in semiconductors, *Springer Series in Solid-State Sciences* **123**, pp. 1–315.

Haug, H., Koch, S.W. (2004). Quantum theory of the optical and electronic properties of semiconductors, *World Scientific*, pp. 1–453.

Hügel, W.A., Heinrich, M.F., Wegener, M., Vu, Q.T., Bányai, L. and Haug, H. (1999). Photon echoes from semiconductor band-to-band continuum transitions in the regime of Coulomb quantum kinetics *Phys. Rev. Lett* **83**, 16, pp. 3313–3316.

Jackson, J.D. (1999), Classical Electrodynamics, *Wiley*, pp. 1–808

Keldysh, L. (1965) Diagramm technique for nonequilibrium processes. *Sov. Phys. JETP* **20**, 4, pp- 1018–1026.

Kirzhnits D. A. (1967). Field Theoretical Methods in Many Body systems, *Pergamon, New York*, pp. 1–394.

Kubo, R. (1957). Statistical-mechanical theory of irreversible processes. *J. Phys. Soc. Japan* **12**, 6, pp. 570–586.

Kubo, R. , Yokota, M. and Nakajima, S. (1957). Statistical Mechanical Theory of Irreversible Processes II. Response to Thermal Disturbance, *J. Phys. Soc. Japan* **12**, 11, pp. 1203–1211.

Kubo, R. , Hasegawa, H. , Hashitsume, N. (1959). Quantum theory of galvanomagnetic effect. I Basic considerations, *J. Phys. Soc. Japan* **14**, 1, pp. 56–74.

Kubo, R., Toda, M. and Hashitsume, N. (1985) Statistical Physics II - Nonequilibrium Statistical Mechanics. *Springer, Berlin, Springer Series in Solid State Sciences* **31** pp. 1–.

Lax, M. (1958). Generalized Mobility Theory, *Phys. Rev.***109**,6, pp. 1921–1926.

Lee, T.D. (1981), Particle Physics and introduction to field theory, *Haarwood Academic, Chur/London/New York*, pp. 1–865.

Louisell, W.H. (1990). Quantum statistical properties of radiation, *Wiley*, pp. 1–528

Luttinger, J. M. (1964). Theory of thermal transport coefficients. *Phys. Rev.* **135**, 6A, pp. 1505–1514.

MacDonald, D. K. C. (1962). Noise and fluctuations: an Introduction. *Wiley, New York, London* pp. 1– .

Mori, H. (1956). Quantum-statistical Theory of Transport Processes, *J. Phys. Soc. Japan*, **11**, 10, pp. 1029–1044.

Nakajima, S. (1958). On Quantum Theory of Transport Phenomena. Steady Diffusion, *Prog. Theor. Phys.* **20**, 6, pp. 948–959.

Nyquist, H. (1928) Thermal agitation of electric charge in conductors. *Phys. Rev.***32**, 1, pp. 110–113.

Pauli, W. (1928) title *Festschrift zum Geburtstage A. Sommerfelds, Leipzig*, pp. 30–

Rammer, J. and Smith, H. (1986), Quantum field-theoretical methods in transport theory of metals. *Rev. Mod. Phys.* **58**, 2, pp. 323–359.

Schäfer, W. , Wegener, M. (2002). Semiconductor Optics and Transport Phenomena, *Springer* , pp. 1–495.

Schmitt, O.M., Tran Thoai, D.B., Bányai, L., Gartner, P., and Haug, H. (2001). Bose-Einstein Condensation quantum kinetics for a gas of interacting excitons. *Phys. Rev. Lett* **86**, 17, pp. 3839–3842.

Titeica, S (1935). Über die Widerstansänderung von Metallen in Magnetfeld. *Annalen der Physik* **22**, 2, pp. 129–161.

van Hove, L. (1955), Quantum-mechanical perturbations giving rise to a statistical transport equation, *Physica* **21**, 5, pp. 517-540.

Weinberg, S. (1996). The Quantum Theory of Fields, *Cambridge University Press*, pp. 1–609.

Wigner, E. P., (1959). Group theory and application to the qantum mechanics of atomic spectra. *Academic Press*, pp. 1–372.

Wiser, N. (1963). Dielectric constant with local field effects included. *Phys. Rev.* **129**, 1, pp. 62–69.

Zwanzig, R. (1960). Ensemble method in the theory of irreversibility *J. Chem. Phys.* **33**, 5, pp.338-1341.

Zwanzig, R. (2001). Nonequilibrium statistical mechanics. *University Press*, pp. 1–222.

Index